1969

LIBRARY
BURROUGHS CORP.

Micropower Circuits

MICROPOWER CIRCUITS

JAMES D. MEINDL
Department of Electrical Engineering
Stanford University

JOHN WILEY & SONS, INC.
New York · London · Sydney · Toronto

Copyright © 1969 by John Wiley & Sons, Inc.

All rights reserved. No part of this book may
be reproduced by any means, nor transmitted,
nor translated into a machine language without
the written permission of the publisher.

Library of Congress Catalog Card Number: 68-28502
SBN 471 59157 2
Printed in the United States of America

Preface

Some of the more familiar devices of consumer, medical, space, and military electronics whose operational effectiveness is markedly dependent on their frugal use of available energy resources are shirt-pocket radio receivers, electronic watches, hearing aids, cardiac pacemakers, communications, meteorological, and navigational satellites, hand-held radio transceivers and radars, and proximity fuses. A primary requirement of the electronic circuits utilized in each of these devices is ultralow power drain. The purpose of this book is to serve as an introduction to the design of semiconductor circuits for which ultralow power drain is a principal performance constraint. These "micropower" circuits generally are identified by transistor quiescent power levels less than approximately one milliwatt.

This book is written for practicing engineers, but the subject of each chapter is developed rigorously from a fundamental viewpoint that can well serve the purpose of advanced undergraduate and graduate study of semiconductor circuits. The treatment presumes that the reader is familiar with the basic principles of semiconductor devices and circuits, which include the more prominent features of integrated circuits.

A design-oriented approach is taken in presenting the material. It assumes that the principal performance constraints of a circuit are dictated by overall system requirements, which cannot be compromised in achieving a micropower circuit. On this basis it becomes clear that a minimum-power design criterion tends to produce a special-purpose rather than a general-purpose circuit. Hence the typical general-purpose integrated circuit is not the focal point of this book. Instead both discrete component and integrated circuits are considered separately and in combination as appropriate from the overriding viewpoint of achieving a micropower circuit.

This book is not intended to be encyclopedic in nature. It is a simple systematic treatment of the more significant topics related to the design of micropower circuits. Physical models of devices especially appropriate for the micropower range are summarized. The influence of bias circuit design on power drain is discussed, and the most common types of linear, nonlinear, and digital micropower circuits are considered in detail. In each chapter a rather complete list of references to the pertinent technical literature is provided.

Perhaps the most salient conclusion a study of micropower circuits can draw is that by judicious design the power drain of virtually all basic types of semiconductor circuits can frequently be reduced by one order of magnitude or more, as compared with common practice.

I am indebted to my colleagues at Stanford University and my former colleagues at the U.S. Army Electronics Command for their aid in the development of this material. In particular, the contributions of Mr. P. H. Hudson and Mr. L. F. Wagner are acknowledged.

JAMES D. MEINDL

Menlo Park, California
August 1968

Contents

I. Device Models   1

    1.1 *p-n* Junction Diodes,   1
    1.2 Junction Transistors,   3
          Large-Signal Static Characteristics,   3
          Large-Signal Dynamic Characteristics,   7
          Small-Signal Characteristics,   10
    1.3 Junction Gate Field Effect Transistors,   14
    1.4 Insulated Gate Field Effect Transistors,   18
    1.5 Tunnel Diodes,   21
    1.6 Passive Devices,   22
    1.7 Parasitic Elements,   26

2. The Quiescent Point   29

    2.1 Device Parameters,   29
    2.2 Specified Power Design,   30
          Emitter Degeneration,   30
          Emitter and Collector Degeneration,   32
          Collector Degeneration and Diode Compensation,   32
          Transistor Compensation,   35
          Emitter Coupled Pair,   35
          Feedback Pair,   35
          Multistage Direct Coupled Circuits,   36
    2.3 Specified Stability Design,   36
    2.4 Effects of Imperfect Element Isolation,   37

3. Low-Frequency Amplifiers 41

3.1 Input Stages, 42
Impedance Matching, 42
Minimum Loading, 43
FET Inputs, 45
3.2 Intermediate Stages, 46
Specified Number of Stages, 46
Specified Total Resistance, 48
3.3 Driver Stages, 50
Basic Design, 52
Low-Distortion Design, 53
Transformer Coupling, 54
3.4 Output Stages, 56
Class B Transformer Coupled Circuits, 57
Class B Complementary Symmetry Circuits, 59
A Monolithic Hearing Aid Amplifier, 59
A Monolithic Acoustic Horn Amplifier, 61
3.5 Class D or Two-State Amplifiers, 61

4. Wideband Amplifiers 67

4.1 Input Stages, 67
4.2 Intermediate Stages, 69
Bipolar Transistor Circuits, 69
FET Circuits, 77
4.3 Output Stages, 81

5. Tuned Amplifiers 84

5.1 The Iterative Stage, 84
Gain, 84
Center Frequency, 86
Bandwidth, 87
Stability, 87
5.2 The Common Emitter Amplifier, 89
5.3 The Cascode Amplifier, 92
5.4 A Monolithic Tuned Amplifier, 96

6. Low-Noise Amplifiers 98

6.1 A Transistor Noise Model, 98
6.2 Noise Figure, 99
6.3 Low-Frequency Amplifiers, 101
Flicker Noise Region, 101
Shot Noise Region, 101

	6.4	High-Frequency Amplifiers, 103	
	6.5	Monolithic Low-Noise Amplifiers, 108	

7. Mixers and Detectors — 110

   7.1 Mixers, 110
       The Equivalent Circuit of a Mixer, 110
       Mixer Noise Figure, 113
       Conversion Gain, 115
   7.2 A Monolithic Mixer, 116
   7.3 Detectors, 116

8. Feedback Amplifiers — 121

   8.1 Sensitivity, 121
       Local Emitter Feedback, 123
       Local Emitter and Collector Feedback, 126
       Alternate Emitter and Collector Feedback Stages, 128
       Multistage Feedback, 129
   8.2 Distortion, 130
   8.3 Bandwidth, 131
   8.4 Stability, 132
   8.5 Quiescent Power, 133

9. Harmonic Oscillators — 134

   9.1 Simplified Design, 134
       Common-Emitter Hartley Oscillator, 135
       Common-Base Hartley Oscillator, 138
   9.2 Generalized Design, 140
   9.3 Micropower Monolithic Oscillators, 143

10. Direct Current Amplifiers — 145

   10.1 DC Gain, 145
   10.2 Common-Mode Rejection, 147
   10.3 Input Offset Voltage, 150
   10.4 Drift, 151
   10.5 Noise, 152
   10.6 Frequency Response, 153
   10.7 A Monolithic Micropower dc Amplifier, 154
   10.8 A Micropower Operational Amplifier, 154

11. Bipolar Transistor Digital Circuits — 157

   11.1 The Basic Inverter Circuit, 157
        Static Behavior, 157
        Dynamic Behavior, 161

Contents

- 11.2 An Iterative Chain of Inverters, 169
- 11.3 Resistor Transistor Logic (RTL), 173
  - Worst Case Analysis, 173
  - Performance Trade-Offs, 180
  - Storage Circuits, 183
- 11.4 Transistor Resistor Logic (TRL), 186
- 11.5 Diode Transistor Logic (DTL), 187
- 11.6 Transistor Transistor Logic (TTL), 189
- 11.7 Emitter Coupled Logic (ECL), 190
- 11.8 Tunnel Device Logic (TDL), 193
- 11.9 Complementary Diode Transistor Logic (CDTL), 196
- 11.10 Pulse Powered Circuits (PPC), 201

12. Field Effect Transistor Digital Circuits   210

- 12.1 The FET Inverter, 210
  - Linear Resistor Loads, 210
  - Depletion Transistor Loads, 216
  - Enhancement Transistor Loads, 217
  - Complementary Transistor Loads, 218
  - Junction Gate Field Effect Transistor Inverters, 222
- 12.2 Logic Circuits 223
- 12.3 Storage Circuits, 226
- 12.4 Multiphase Dynamic Circuits, 229
  - Two-Phase Circuits, 231
  - Four-Phase Circuits, 233

13. Applications   238

- 13.1 Portable Equipment, 238
- 13.2 Aerospace Electronics, 241
- 13.3 Biomedical Electronics, 243

List of Symbols   247

Index   259

Micropower Circuits

Chapter 1

Device Models

The design of a micropower circuit involves (a) the selection of appropriate devices, (b) the choice of a suitable circuit configuration, and (c) the minimization of the total operating power of the circuit considering its primary performance constraints. A necessary feature of the design process is the adoption of a model for each device that adequately describes its salient characteristics in the micropower range. This chapter considers static and dynamic models of active as well as passive devices in the micropower range.

1.1 p-n JUNCTION DIODES

Under the condition of low-level injection, the total current in an ideal p-n junction diode is a bulk diffusion current that may be written as [1,2]

$$I = I_S(\epsilon^{qV/kT} - 1), \tag{1.1}$$

where

$$I_S = qA_j\left(\frac{D_n n_{p0}}{L_n} + \frac{D_p P_{n0}}{L_p}\right). \tag{1.2}$$

Two salient features of this current-voltage relationship are a saturation current I_S under large reverse bias and a simple exponential dependence of the forward current on the applied bias for V greater than several kT/q. It has been observed that the measured current-voltage characteristics of certain p-n junctions deviate from the ideal diffusion model [3,4]. In the case of silicon diffused planar *p-n* junctions, this deviation is of particular interest for micropower devices.

The current flowing in a nonideal *p-n* junction may be divided into four components according to the location of the recombination and generation of electrons and holes. These are [4] (a) the ideal diffusion current or the bulk recombination-generation (*r-g*) current of the neutral region given by (1.1), (b) the transition region bulk *r-g* current, (c) the transition region surface *r-g* current, and (d) the neutral region surface *r-g* current including surface channel current. The nonideal or excess components of junction current may be written empirically as

$$I = I_X \epsilon^{qV/nkT}, \qquad (1.3a)$$

where

$$I_X \simeq \tfrac{1}{2} q W_t A_j \frac{n_i}{\tau_0}. \qquad (1.3b)$$

$V > 4(kT/q)$ and $1 < n < 2$ if surface channels are not present. The value of n is greater than 2 for surface channel current. For reverse junction voltages $-I_X(V)$ cannot saturate since the generation current depends on the width of the transition region, which varies with applied voltage. For many silicon *p-n* junctions $I_X \gg I_S$ so that total junction current is dominated by excess current for a reverse-bias as well as at relatively small forward-bias voltages. Consequently the nonideal behavior of a silicon junction is of particular importance at the low operating current levels of micropower circuits.

An empirically derived composite of (1.1) and (1.3),

$$I = I_J \epsilon^{qV/nkT}, \qquad V > \frac{4kT}{q}, \qquad (1.4)$$

introduces the effect of *r-g* current on the small-signal forward conductance of the junction:

$$\frac{\partial I}{\partial V} = g = \frac{qI}{nkT}. \qquad (1.5)$$

The total small-signal capacitance of a *p-n* junction C_j may be divided into three components, which are [5] (a) C_d, the minority carrier diffusion capacitance outside the transition region; (b) C_{ti}, the transition space charge capacitance caused by immobile ionized impurities; and (c) C_{tc}, the transition carrier capacitance resulting from free carriers in the transition region. Under a reverse bias, C_{ti} dominates the total junction capacitance, which is accurately described for a linearly graded junction by

$$C_j \simeq C_{ti} = A_j \left(\frac{q a_t \epsilon^2}{12} \right)^{1/3} (\phi - V)^{-1/3} \qquad (1.6)$$

for $(\phi - V)$ less than a few kT/q. In a p-n^+ diode the diffusion capacitance is

$$C_d \simeq q \frac{n_i^2}{a_t} \frac{q}{kT} A_j \epsilon^{qV/kT}, \qquad L_n \ll W_t \tag{1.7}$$

or

$$C_d \simeq q \frac{n_i^2}{a_t} \frac{q}{kT} \frac{2L_n}{W_t} A_j \epsilon^{qV/kT}, \qquad L_n \gg W_t, \tag{1.8}$$

where L_n is the minority carrier diffusion length and W_t is the width of the linearly graded transition region (see List of Symbols). Since [5]

$$C_{tc} \simeq \frac{qn_i}{6} \frac{q}{kT} W_t A_j \epsilon^{qV/nkT} \tag{1.9}$$

is large compared with C_{ti} for moderate and large forward voltages, comparing (1.7) and (1.8) with (1.9) indicates that C_{tc} may dominate C_j at moderate forward voltages in materials with large energy gap, small intrinsic carrier concentration n_i, low lifetime, and large concentration gradient at the junction. At the low operating current levels of silicon micropower devices, C_{tc} may dominate the capacitance of a junction under forward bias.

1.2 JUNCTION TRANSISTORS

The exponential current-voltage characteristic of a *p-n* junction (1.4) prescribes that micropower operation of transistors be achieved through orders-of-magnitude reductions in junction currents and almost incremental reductions in junction voltages compared with typical low-power operating conditions. Consequently the essential qualities of a micropower transistor are (a) junction reverse currents that are small compared with minimum operating currents, (b) usable device current gains at minimum operating currents, and (c) minimum junction capacitances in order to enhance frequency response.

Large-Signal Static Characteristics

The collector current-voltage characteristics of a typical silicon diffused planar micropower transistor are illustrated in Figure 1.1. Logarithmic scales are used to display the striking similarities in the static behavior of the transistor over several decades of current and voltage. The salient features of the collector characteristics can be derived from the simple two-diode model of an idealized transistor shown in Figure 1.2a. The static equations

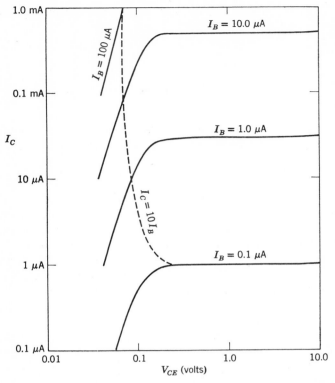

Figure 1.1 Transistor collector current-voltage characteristics.

of this model are [2,6]

$$I_E = -I_{EBS}(\epsilon^{qV_{BE}/kT} - 1) + \alpha_I I_{CBS}(\epsilon^{-qV_{CB}/kT} - 1) \quad (1.10a)$$

and

$$I_C = \alpha_N I_{EBS}(\epsilon^{qV_{BE}/kT} - 1) - I_{CBS}(\epsilon^{-qV_{CB}/kT} - 1). \quad (1.10b)$$

The dynamic model of Figure 1.2b is discussed in the following subsection.

The excess junction currents caused by recombination and generation in the transition regions and on the surface of the base region [4,7] can be added to (1.10) to give

$$I_E = -2I_{EX} \sinh\left(\frac{qV_{BE}}{nkT}\right) - I_{EBS}(\epsilon^{qV_{BE}/kT} - 1)$$
$$+ \alpha_I I_{CBS}(\epsilon^{-qV_{CB}/kT} - 1) \quad (1.11a)$$

and

$$I_C = -2I_{CX} \sinh\left(\frac{-qV_{CB}}{nkT}\right) + \alpha_N I_{EBS}(\epsilon^{qV_{BE}/kT} - 1)$$

$$- I_{CBS}(\epsilon^{-qV_{CB}/kT} - 1). \quad (1.11b)$$

Since $I_E + I_C + I_B = 0$, (1.11) give

$$I_B = (1 - \alpha_N)I_{EBS}(\epsilon^{qV_{BE}/kT} - 1) + (1 - \alpha_I)I_{CBS}(\epsilon^{-qV_{CB}/kT} - 1)$$

$$+ 2I_{EX} \sinh\left(\frac{qV_{BE}}{nkT}\right) + 2I_{CX} \sinh\left(\frac{-qV_{CB}}{nkT}\right). \quad (1.12)$$

For $V_{BE} > 4(kT/q)$ and $V_{CB} = 0$,

$$I_C = \alpha_N I_{EBS}\epsilon^{qV_{BE}/kT} \quad \text{and} \quad I_B = (1 - \alpha_N)I_{EBS}\epsilon^{qV_{BE}/kT} + I_{EX}\epsilon^{qV_{BE}/nkT}. \quad (1.13)$$

Figure 1.3 illustrates these equations for a typical micropower transistor.

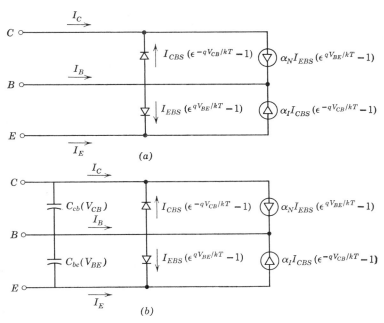

Figure 1.2 Idealized large signal transistor models: (a) static (b) dynamic.

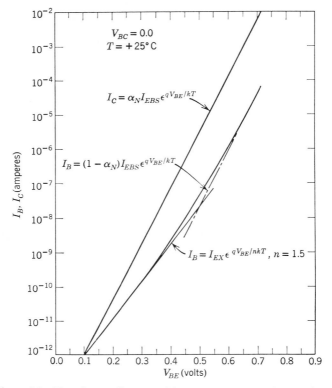

Figure 1.3 Transistor collector and base currents versus base-emitter voltage.

The fall-off of current gain,

$$h_{FE} = \frac{I_C - I_{CEO}}{I_B} \simeq \frac{I_C}{I_B}$$

$$= \left[\frac{1 - \alpha_N}{\alpha_N} + \frac{I_{EX}}{(\alpha_N I_{ES})^{1/n}} I_C^{-(n-1)/n} \right]^{-1} \quad (1.14)$$

with decreasing I_C observed in virtually all silicon micropower transistors is described by (1.14) as a consequence of the excess components of emitter junction current [4,8,9]. Tunnel currents in surface inversion layers have been reported as a further source of degradation of h_{FE} [10]. In addition, silicon junctions that exhibit virtually ideal $\exp qV/kT$ behavior as well as an essentially constant value of h_{FE} over many decades of I_C have been observed [11].

Since the low frequency small signal current gain is $h_{fe0} = \dfrac{\partial I_C}{\partial I_B} \simeq \dfrac{\Delta I_C}{\Delta I_B}$, (1.14) gives

$$\frac{h_{fe0}}{h_{FE}} = \frac{\beta_N + (I_N/I_C)^{(n-1)/n}}{\beta_N + \dfrac{1}{n}(I_N/I_C)^{(n-1)/n}}$$

$$\simeq n, \quad \frac{I_N}{I_C} \gg 1$$

$$\simeq 1, \quad \frac{I_N}{I_C} \ll 1 \tag{1.15a}$$

with $\beta_N = \alpha_N/(1 - \alpha_N)$ and

$$I_N = \frac{I_{EX}^{n/(n-1)}}{(\alpha_N I_{ES})^{1/(n-1)}}. \tag{1.15b}$$

This result indicates that an increase in the excess current coefficient of the emitter junction, I_{EX}, relative to the saturation current I_{EBS} tends to enlarge h_{fe0} in comparison to h_{FE}. An increase in the junction ideality factor n has a similar effect. As I_C increases, the excess current becomes less significant so that h_{fe0} and h_{FE} are more nearly equal.

Large-Signal Dynamic Characteristics

The large-signal static equations of the transistor (1.10) can be converted to a more convenient form [6,12],

$$I_E = -\alpha_I I_C - I_{EBO}(\epsilon^{qV_{BE}/nkT} - 1) \tag{1.16a}$$

and

$$I_C = -\alpha_N I_E - I_{CBO}(\epsilon^{-qV_{CB}/nkT} - 1) \tag{1.16b}$$

for a discussion of both large-signal and small-signal dynamic models for the micropower range. In (1.16),

$$I_{EBO} = (1 - \alpha_N \alpha_I)I_{EBS}, \quad I_{CBO} = (1 - \alpha_N \alpha_I)I_{CBS},$$

and

$$\alpha_N I_{EBO} = \alpha_I I_{CBO}. \tag{1.17}$$

The large-signal dynamic behavior of the transistor [13,14] can be described by simply adding the diffusion and transition region capacitances to the model described by (1.16) as illustrated in Figure 1.4a. The complete model of Figure 1.4a can be converted to the form of Figure 1.4b for operation in the active region or of Figure 1.4c for the cutoff region. In addition to the models

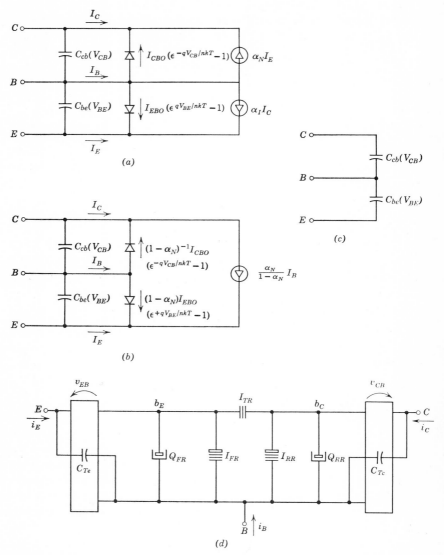

Figure 1.4 (a) Generalized large-signal dynamic transistor model, (b) active region model, (c) cutoff region model, (d) a lumped model.

of Figure 1.4, the charge control model [15] or the lumped model [16] of Figure 1.4d can be used to describe transistor large-signal dynamic behavior. The differential equations of the lumped model are

$$i_E = -(I_{TR} + I_{FR})b_E - Q_{FR}\frac{db_E}{dt} + C_{Te}\frac{dv_{BE}}{dt} + I_{TR}b_C \quad (1.18a)$$

and

$$i_C = I_{TR}b_E - (I_{TR} + I_{RR})b_C - Q_{RR}\frac{db_C}{dt} - C_{Tc}\frac{dv_{BC}}{dt}, \quad (1.18b)$$

where b_E = the ratio of excess minority carrier density to equilibrium minority carrier density at the emitter = $\exp(qV_{BE}/kT) - 1$;

b_C = relative excess minority carrier density at the collector = $\exp(qV_{BC}/kT) - 1$;

$I_{TR}b_E$ = portion of emitter forward current which is collected; I_{TR} is a reference transport current;

$I_{FR}b_E$ = portion of emitter forward current that is lost through recombination; I_{FR} is a forward reference recombination current;

$Q_{FR}b_E$ = stored minority carrier charge associated with emitter forward current; Q_{FR} is a forward reference charge;

$I_{TR}b_C$ = portion of collector forward current that is collected;

$I_{RR}b_C$ = portion of collector forward current that is lost through recombination; I_{RR} is a reverse reference recombination current;

$Q_{RR}b_C$ = stored minority carrier charge associated with collector forward current; Q_{RR} is a reverse reference charge;

C_{Te} = emitter junction transition region capacitance; and

C_{Tc} = collector junction transition region capacitance.

The transistor capacitances $C_{be}(V_{BE})$ and $C_{cb}(V_{CB})$ (Figure 1.5) are largely composed of transition region space charge and carrier capacitances in the micropower range. These capacitances dominate the switching speed (and gain-bandwidth product) of the transistor [17,18]. The capacitance limited delay-time, rise-time, and fall-time increase in proportion to the reduction of operating current. Storage time, however, does not increase with lower currents since the amount of excess stored charge in a saturated transistor is reduced in proportion to the currents that must remove the charge. Consequently, storage time is usually a negligible component of the switching time of a micropower transistor. A pronounced property of a micropower transistor inverter is that the circuit fall time must be divided into two intervals [17,19]. During the first interval the transistor is active; during the second it is cutoff.

The influence of r-g currents can be incorporated in the model of Figure 1.4a by simply treating the parameters, α_N, α_I, I_{EBO}, I_{CBO}, C_{be}, C_{cb}, and n, as

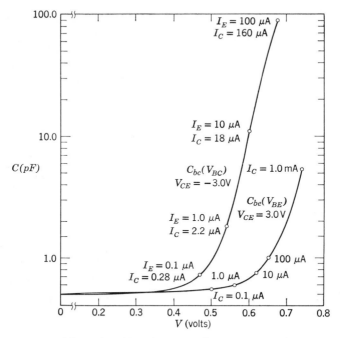

Figure 1.5 Transistor capacitances versus voltage.

measurable quantities that include the effects of *r-g* currents. Since these parameters vary with the dc operating point, average values for a given switching interval can be used.

Small-Signal Characteristics

In the micropower range, the hybrid pi model of the transistor provides an appropriate small-signal equivalent circuit [20]. A complete hybrid pi model [21,22] is illustrated in Figure 1.6a and a simplified version, adequate in many micropower applications, is illustrated in Figure 1.6b. The principal assumption on which the simplification is based may be expressed as

$$r_b \ll \left| \frac{r_{be}}{1 + j\omega r_{be} C_{be}} \right|, \tag{1.19}$$

which is rarely invalid at relatively low values of I_C since the base-to-emitter diffusion resistance, r_{be}, is inversely proportional to I_C whereas the series base resistance, r_b, and base-to-emitter capacitance, C_{be}, are extremely small for the highest-frequency transistors of interest. The capacitances, C_{be} and C_c, include both diffusion and transition region contributions. The collector capacitance directly under the emitter is represented by C_{c1}; C_{c2} represents the remaining collector-to-base capacitance. The resistances r_c and r_{ce}, which result from base-width modulation, are typically quite large compared with r_{be}.

Because of the high impedance levels of micropower small-signal circuits as well as the use of transistors with extremely small junction capacitances ($C_{be} \sim 5$ pF), stray capacitances associated with isolation junctions in a monolithic integrated circuit, with device packages, and with interstage wiring can drastically reduce the frequency response of a circuit. Consequently, a circuit parameter C_s representing stray capacitance should be included explicitly in the analysis of many micropower circuits, as Figure 1.7 illustrates.

From the model of Figure 1.7 the gain-bandwidth product of the transistor can be calculated as

$$f_T = \frac{g_m}{2\pi(C_{be} + C_c + C_{s1} + C_{s2})}. \tag{1.20}$$

This equation provides an accurate description of the curves of Figure 1.8

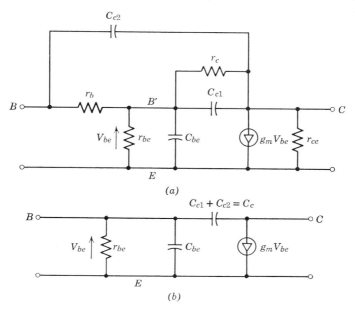

Figure 1.6 Transistor small-signal circuit models: (a) complete hybrid pi model; (b) simplified hybrid pi model.

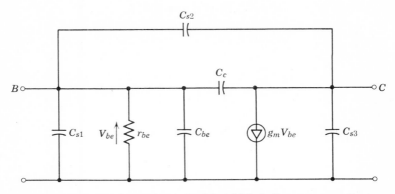

Figure 1.7 Simplified small-signal circuit model including stray capacitances.

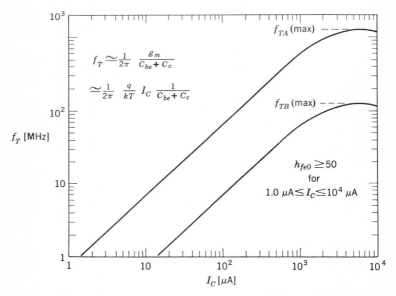

Figure 1.8 Transistor gain-bandwidth product versus quiescent collector current. $V_{CE} = 3.0V$ for transistors A and B.

for $f_T < 0.5f_T$ (max) under the conditions that

$$g_m = \frac{qI_C}{kT}, \quad r_{be} = \frac{h_{fe0}}{g_m}, \quad (1.21)$$

and C_{be} is essentially independent of I_C (Figure 1.5). The effects of r-g currents are reflected in the measured values of h_{fe0}, C_{be}, and C_c.

A principal problem associated with minimizing the quiescent power of a linear micropower circuit is the selection of an active device [23,24,25]. A guide for device selection is as follows: Where other ratings permit, select the transistor with the maximum gain-bandwidth product or frequency response at a given operating current. Often a large quiescent power saving can be achieved by using an extremely small geometry microwave transistor at a frequency well below its maximum capability; for example, suppose that it has been established that the circuit under consideration requires a transistor current gain-bandwidth product $f_T = 120$ MHz. Figure 1.8 indicates that both device A with f_{TA} (max) = 1200 MHz at $I_{CA} = 6.0$ mA and device B with f_{TB} (max) = 120 MHz at $I_{CB} = 6.0$ mA can provide $f_T = 120$ MHz. However, the required collector currents are $I_{CA} = 0.20$ mA and $I_{CB} = 6.0$ mA. In this case the use of device A would permit a quiescent power $\frac{1}{30}$ as large as the power that device B would require.

From (1.6)–(1.9) it is evident that all components of the capacitance of a junction are directly proportional to the area A_j of the junction. Consequently (1.20) indicates that gain-bandwidth product is inversely proportional to junction area in the micropower range assuming negligible parasitics. Disregarding sidewall effects, for a planar emitter junction capacitance per unit area of 1 pF/mil², subject to the assumption $C_{be} = C_c$, the following data indicate the dependence of f_T on A_j (emitter).

A_j (emitter)	$f_T(I_C = 10\ \mu A)$	$I_C(f_T = 10$ MHz)
1 mil × 1 mil = 625 microns²	31.8 MHz	3.1 μA
0.1 mil × 0.1 mil = 6.25 microns²	3180 MHz	0.031 μA

$$(C_{be}/A_j = 1\ \text{pF/mil}^2 = 0.016\ \text{pF/micron}^2; C_{be} = C_c)$$

$$(1\ \text{mil} = 25\ \text{microns})$$

Metal bonding pad capacitances of 0.01–0.06 pF/mil² alone are sufficient to mask the capacitance of the 6.25-micron² emitter transistor and to degrade the performance of the 625-micron² device. Therefore it is clear that to realize the ultimate micropower potential of the transistor [26], it must be embedded in a compatible integrated circuit environment whose interconnections and passive elements [24] do not negate the potential of the transistor.

1.3 JUNCTION GATE FIELD EFFECT TRANSISTORS

The drain current-voltage characteristics of a typical silicon diffused planar junction gate field effect transistor (JGFET) suitable for micropower operation are illustrated in Figure 1.9. For an abrupt gate junction, the static behavior in the triode region before pinch-off is described by the equation [27,28]

$$I_D = \frac{3I_{DSS}}{V_P} \left\{ (V_{DS} - V_{GS}) \left[1 - \frac{2}{3} \left(\frac{V_{DS} - V_{GS}}{V_P} \right)^{1/2} \right] \right.$$
$$\left. - (V_S - V_{GS}) \left[1 - \frac{2}{3} \left(\frac{V_S - V_{GS}}{V_P} \right)^{1/2} \right] \right\} \quad (1.22)$$

for an *n* channel device where the pinch-off voltage

$$V_P = \frac{\sigma a_c^2}{2\mu_c \epsilon} \quad \text{and} \quad I_{DSS} = \frac{2\sigma a_c}{L_c} \frac{V_P}{3} W_c. \quad (1.23)$$

At pinch-off ($V_{DS} - V_{GS} = V_P$) and for $V_S = 0$, (1.22) yields

$$I_D = I_{DSS} \left\{ 1 + \frac{V_{GS}}{V_P} \left[3 - 2 \left(\frac{-V_{GS}}{V_P} \right)^{1/2} \right] \right\}, \quad (1.24)$$

which provides a useful approximation for the static behavior of the JGFET in the pinch-off region. The nonzero slope of the drain current-voltage

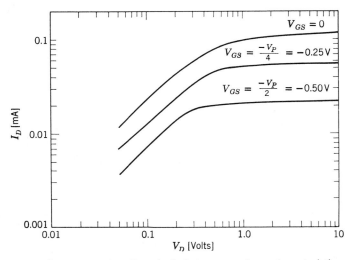

Figure 1.9 JGFET static drain current-voltage characteristics.

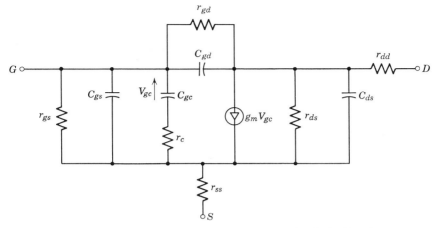

Figure 1.10 Complete JGFET small-signal model in pinch-off region.

characteristics in the pinch-off region is the principal feature of the device behavior not reflected in (1.24). A rather accurate approximation for the uniform channel case of (1.22), as well as virtually any channel impurity profile [29,30,31], is given by

$$I_D = \frac{2I_{DSS}}{V_P} V_{DS}\left(1 + \frac{V_{GS} - V_{DS}/2}{V_P}\right) \quad (1.25)$$

in the triode region and

$$I_D = I_{DSS}\left(1 + \frac{V_{GS}}{V_P}\right)^2 \quad (1.26)$$

in the pinch-off region.

Figure 1.10 illustrates a complete grounded source small-signal circuit model of the JGFET in the pinch-off region [32]. From (1.24) the transconductance is

$$\frac{\partial I_D}{\partial V_{GS}} = g_m = \frac{3I_{DSS}}{V_P}\left[1 - \left(\frac{-V_{GS}}{V_P}\right)^{1/2}\right], \quad (1.27)$$

where

$$g_{m0} = \frac{3I_{DSS}}{V_P} \quad (1.28)$$

is the maximum transconductance as well as the unmodulated channel conductance. In the pinch-off region the abrupt junction gate-to-channel capacitance [32, 33] is

$$C_{gc} = C_0 \frac{1 + \sqrt{-V_{GS}/V_P}}{(1 + 2\sqrt{-V_{GS}/V_P})^2}, \quad (1.29)$$

where

$$C_0 = 6\epsilon \frac{L_c}{a_c} W_c \quad (1.30)$$

and the effective channel resistance [33] is

$$r_c = \frac{9}{35} \frac{1}{g_{m0}} \frac{[1 + 5\sqrt{-V_{GS}/V_P} - \frac{10}{3}(V_{GS}/V_P)]}{[1 + (V_{GS}/V_P)](1 + 2\sqrt{-V_{GS}/V_P})}. \quad (1.31)$$

The transconductance cutoff frequency is

$$f_c = \frac{1}{2\pi r_c C_{gc}}, \quad (1.32)$$

which is approximately a maximum [33]

$$f_{c0} = \frac{g_{m0}}{2\pi(\frac{9}{35})C_0}$$

$$= \frac{1}{2\pi}\left(\frac{35}{27}\right)\frac{\sigma}{\epsilon}\left(\frac{a_c}{L_c}\right)^2 \quad (1.33)$$

for $V_G = 0$.

The capacitances in the model of Figure 1.10 are functions of the gate and drain potentials. For example, in the JGFET of Figure 1.11 with gate 1 and gate 2 connected, as well as the source and substrate at ground potential, note the following:

1. C_{gc} is approximately the sum of the gate-to-channel junction capacitances in regions bc and fg.

2. C_{gs} is approximately the sum of the junction capacitances in regions cd, gh, and ij.

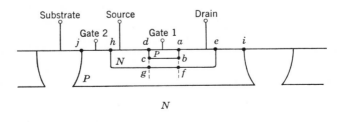

Figure 1.11 Typical diffused JGFET structure within an integrated circuit.

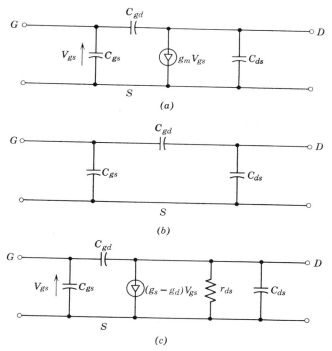

Figure 1.12 Simplified JGFET small-signal models for (a) pinch-off region, (b) cutoff region, (c) triode region.

3. C_{gd} is approximately the sum of the junction capacitances in regions *ab* and *ef*.

The elements C_{ds} and r_{ds} of the model of Figure 1.10 depend on the properties of the pinched-off region of the channel [27,29]. The resistances r_{gd} and r_{gs} are those of junctions under reverse bias, and r_{dd} and r_{ss} are bulk resistances.

If the operating frequency of the JGFET is much less than f_c, the complete pinch-off region small-signal model of Figure 1.10 can be simplified as shown in Figure 1.12a. For operation in the cutoff region where $V_{DS} - V_{GS} > V_P$, $V_S - V_{GS} > V_P$ and $I_D \simeq 0$, the model of Figure 1.12b is valid. A simplified model for the triode region is illustrated in Figure 1.12c. In this model

$$\frac{\partial I_D}{\partial V_{GS}} = g_{m0}\left[1 - \left(\frac{V_S - V_{GS}}{V_P}\right)^{1/2}\right] - g_{m0}\left[1 - \left(\frac{V_{DS} - V_{GS}}{V_P}\right)^{1/2}\right]$$
$$= g_s - g_d, \qquad (1.34)$$

and

$$\frac{\partial I_D}{\partial V_{DS}} = g_d = \frac{1}{r_{ds}} \tag{1.35}$$

can be derived from (1.22).

In a bipolar transistor circuit, the only means for adjusting device transconductance $g_m = qI_C/kT$ to the particular value required for a minimum power design is to adjust the quiescent collector current I_C. This technique is convenient and quite effective over a g_m (and I_C) range of many orders of magnitude. In a JGFET, however, as (1.27) indicates, control of g_m by adjustment of the gate voltage is reasonably effective only for a g_m range of about one order of magnitude. Larger changes in g_m must be made by changing $g_{m0} = 3I_{DSS}/V_P$, which requires a change in the device geometry, most frequently the channel width W. Ideally, W should be a design variable for JGFET (as well as IGFET) micropower circuits.

1.4 INSULATED GATE FIELD EFFECT TRANSISTORS

The insulated gate field effect transistor (IGFET) is a useful device in both linear and digital micropower circuits. The drain current voltage characteristics of a typical depletion-type IGFET suitable for micropower circuits are illustrated in Figure 1.13a; the characteristics of an enhancement-type IGFET are shown in Figure 1.13b. The static drain characteristics are described approximately by [34–37]

$$I_D = 2K[(V_{GS} - V_P)V_{DS} - \tfrac{1}{2}V_{DS}^2] \tag{1.36}$$

for an n channel device in the triode region with

$$K = \frac{\mu_c \epsilon_{ox} W_c}{2T_{ox}L_c}, \tag{1.37}$$

$V_S = 0$ and V_P the pinch-off or threshold voltage defined as the gate-to-channel voltage for which the mobile channel charge goes to zero. Pinch-off voltage may be positive, negative, or zero for the IGFET. At pinch-off, $V_{DS} = V_{GS} - V_P$, (1.36) yields

$$I_D = K(V_{GS} - V_P)^2, \tag{1.38}$$

which is comparable to (1.24) and (1.26) for the JGFET.

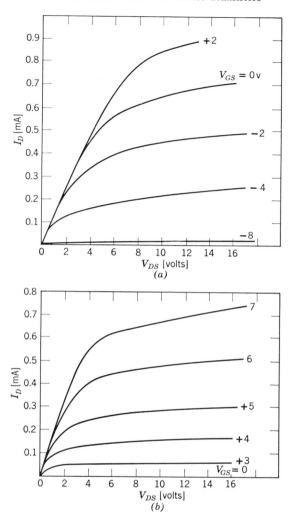

Figure 1.13 (a) Depletion-type IGFET drain characteristics; (b) Enhancement-type IGFET drain characteristics.

Disregarding the series bulk resistances r_{dd} and r_{ss}, the small-signal model of Figure 1.10 provides an adequate description of the IGFET in the pinch-off region. From (1.38) the IGFET transconductance is

$$\frac{\partial I_D}{\partial V_{GS}} = g_m = 2K(V_{GS} - V_P). \tag{1.39}$$

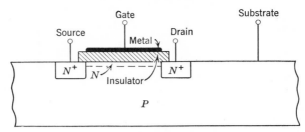

Figure 1.14 Typical IGFET structure.

The transconductance cutoff frequency

$$f_c = \frac{1}{2\pi r_c C_{gc}} \approx \frac{g_m}{2\pi C_{gc}} \qquad (1.40)$$

depends on the active channel resistance r_c and capacitance C_{gc}. The capacitances C_{gc}, C_{gs}, C_{gd}, and C_{ds} are functions of the gate, drain, and substrate potentials; for example, note the following in the IGFET of Figure 1.14 with the source and substrate common:

1. C_{gc} is the gate-to-channel capacitance.
2. C_{gs} is the direct gate-to-source capacitance including the result of the portion of the gate electrode that overlaps the source.
3. C_{gd} is typically the result of direct channel-to-drain feedback capacitance through the drain-channel depletion region plus gate electrode overlap.
4. C_{ds} is largely the drain-to-substrate junction capacitance.

The drain resistance r_{ds} depends on the properties of the pinch-off depletion region, and r_{gs} and r_{gd} are extremely large leakage resistances of the input metal-oxide-semiconductor capacitor.

If the operating frequency of the IGFET is much less than f_c, the simplified models of Figure 1.11 may be used to describe the device. In the triode region

$$\frac{\partial I_D}{\partial V_{GS}} = 2KV_{DS} = 2K[V_{GS} - (V_{GS} - V_{DS})]$$

$$= g_s - g_d \qquad (1.41)$$

and

$$\frac{\partial I_D}{\partial V_{DS}} = 2K[(V_{GS} - V_{DS}) - V_P]$$

$$= \frac{1}{r_{ds}} \qquad (1.42)$$

can be derived from (1.36).

1.5 TUNNEL DIODES

By appropriate adjustment of the peak current, I_P, during the manufacturing process, tunnel diodes suitable for micropower operation can be produced. The static current-voltage characteristic of a germanium micropower tunnel diode is illustrated in Figure 1.15 [38]. Although a theoretical expression for this characteristic has been developed [39], it is often convenient to use polynomial, piecewise linear, or exponential functions to represent the current-voltage characteristics [40]. An exponential representation [41] is

$$I = I_1 + I_2$$
$$= AV\epsilon^{-aV} + B(\epsilon^{bV} - 1), \tag{1.43}$$

where the first term approximates the tunneling currents and the second term gives the normal forward-biased diode current. In (1.43), A, a, B, and b are empirically determined constants.

A complete small-signal model for the tunnel-diode in the negative resistance region is illustrated in Figure 1.16a [40]. For operation well below

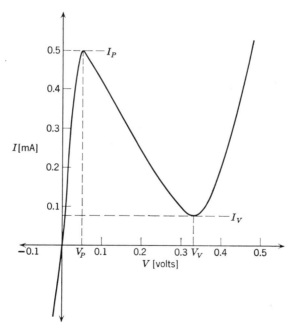

Figure 1.15 Tunnel-diode static current-voltage characteristic.

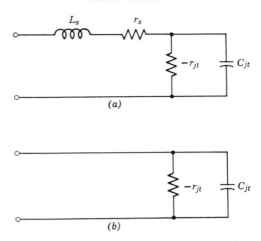

Figure 1.16 (a) Complete tunnel-diode small-signal model; (b) simplified tunnel-diode small-signal model.

the maximum frequency of oscillation,

$$f_{\max} = \frac{1}{2\pi r_{jt} C_{jt}} \left(\frac{r_{jt}}{r_s} - 1 \right)^{\!1/2}, \tag{1.44}$$

the series resistance r_s and package inductance L_s of the diode may be neglected, and the simplified model of Figure 1.16b provides an accurate description of the device behavior. The cutoff frequency of the simplified model is

$$f_{ct} = \frac{1}{2\pi r_{jt} C_{jt}}, \tag{1.45}$$

where $-r_{jt} = (\partial I/\partial V)^{-1}$ is the negative resistance of the junction and C_{jt} is the abrupt junction capacitance. Tunnel rectifiers or backward diodes—tunnel-diodes for which $I_P/I_V \simeq 1$—exhibit relatively low knee voltages in their forward direction and high currents in their reverse direction (Figure 1.15).

1.6 PASSIVE DEVICES

The most pronounced difference between the passive devices used in typical low-power circuits and those used in micropower circuits is the relatively large-value resistors that must accompany the smaller quiescent

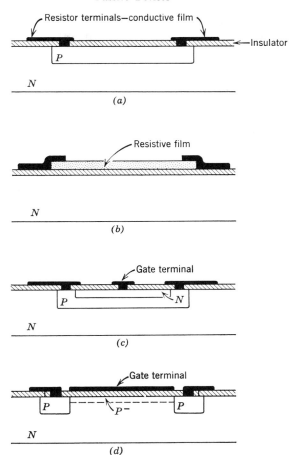

Figure 1.17 Monolithic resistor structures: (a) diffused layer resistor; (b) thin film resistor; (c) junction gate field effect resistor; (d) insulated gate field effect resistor.

currents of micropower circuits. Particularly in monolithic integrated circuits, the parasitic capacitances associated with large-value resistors can produce a very significant effect on circuit performance.

High-value, low-power thallium oxide discrete resistors have been developed specifically for use in the micropower range [42]. Except for their small size and low power rating, these devices are quite similar to normal high-value discrete resistors. The principal reactive element associated with discrete resistors in micropower circuits is a small parasitic shunt capacitance, usually well below one picofarad.

Figure 1.17 illustrates four techniques that are used to obtain resistors in monolithic integrated circuits. The resistance of these structures may be expressed as

$$R = \rho_s \frac{L_r}{W_r}, \qquad (1.46)$$

where ρ_s is the sheet resistivity of the resistive region in ohms per square, L_r the length, and W_r the width of the resistive path. The diffused layer resistor [43,44] of Figure 1.17a is the most commonly used in monolithic circuits. In the micropower range the large values required tend to dominate the associated design problem because of the following factors:

1. The sheet resistivity ρ_s is limited in value since, typically, it must satisfy the design constraints of a transistor base diffusion.
2. The resistor width W_r is limited to a minimum value determined by photolithographic and etching techniques.
3. Resistor length L_r is increased only at the expense of enlarging both the substrate area occupied by the resistor and its parasitic capacitance. In principle, the limitation of ρ_s is removed in the case of the thin film resistor [43,45] of Figure 1.17b. However, in practice the added processing necessary for compatibly depositing thin films of high sheet resistivity poses serious technological and economic problems. To the extent that these problems can be overcome, the thin film resistor, with its relatively small parasitic capacitance compared with the diffused layer resistor, is a most desirable passive element for micropower circuits.

The field effect resistors of Figure 1.17c and 1.17d can be obtained, in essence, by operating field effect transistors in the triode region. They are desirable chiefly because they provide a readily available technique for achieving large sheet resistivities. These three terminal resistors provide typical sheet resistivities of 5000–50,000 ohms per square compared with 200–2000 ohms/sq for thin film resistors and 100–400 ohms/sq for diffused layer resistors. The bipolar transistor base and emitter diffusions typically used to form the resistor of Figure 1.17c [46] result in relatively high resistor tolerances, parasitic capacitances per unit area, temperature sensitivity, and reverse junction current densities compared with the diffused layer resistor of Figure 1.17a. With an additional supply voltage to bias the gate [47], the insulated gate field effect resistor of Figure 1.17d is quite useful in micropower circuits.

A series of models applicable to the resistors of Figure 1.17a and 1.17b is illustrated in Figure 1.18. Here R is the total value of the resistance and C the total parasitic capacitance. The manufacturing tolerance, temperature coefficient, and voltage coefficient of both the resistor and its parasitic

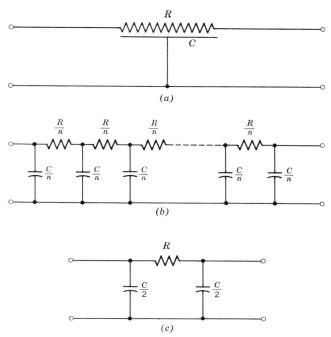

Figure 1.18 Monolithic resistor models: (*a*) distributed model; (*b*) complete lumped model; (*c*) simplified lumped model.

capacitance should be taken into account in a micropower circuit design. In addition, the frequency response of the resistor, whose cutoff frequency is given by

$$f_{cr} \simeq \frac{1}{\pi RC} \tag{1.47}$$

from Figure 1.18*c*, must be recognized. Circuit models for the resistors of Figure 1.17*c* and 1.17*d* may be derived from Figures 1.10, 1.12, and 1.18.

Because of the larger impedances of micropower circuits, it becomes feasible, if not advantageous, to use monolithic capacitors more frequently than in typical low-power integrated circuits. A *p-n* junction monolithic capacitor structure is illustrated in Figure 1.19*a* and a simplified circuit model in Figure 1.19*b* [48]. The physical structure of this capacitor tends to limit the ratio of the desired capacitance to the parasitic capacitance, C_1/C_2, to values from 5 to 10 depending on the applied reverse junction voltages. The parasitic capacitance must be accounted for in the analysis and design of a micropower circuit.

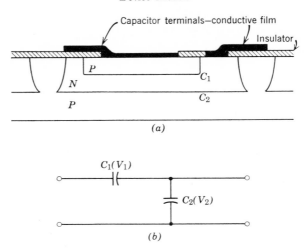

Figure 1.19 *p-n*-junction capacitor: (*a*) capacitor structure; (*b*) simplified circuit model.

1.7 PARASITIC ELEMENTS

Using currently available devices, micropower circuits are designed by reducing operating currents while sustaining operating voltages compared with typical low-power circuits. The larger circuit impedances that result, coupled with the extremely small terminal capacitances of micropower transistors, provide a setting in which typically negligible parasitic capacitances can cause a pronounced effect on circuit behavior (e.g., on the quiescent current required for a specified amplifier bandwidth). Consequently, the importance of isolation junction capacitances, package capacitances, wiring capacitances, and other forms of stray capacitance must be presumed in micropower circuits.

Careful design of the layout and packaging of a circuit is rather critical. Microminiature construction is highly desirable not only to reduce packaging, wiring, and other stray parasitics but also to provide added opportunity for shielding. The importance of the geometrical configuration of a micropower circuit bears some comparison to this feature of a microwave circuit. In general, the penalty for disregarding the detailed nature of the environment of a micropower circuit is larger power consumption.

REFERENCES

[1] W. Shockley, "The Theory of *p-n* Junctions in Semiconductors and *p-n* Junction Transistors," *Bell System Tech. J.*, **28**, 435–489 (July 1949).

[2] J. L. Moll, *Physics of Semiconductors*, McGraw-Hill, New York, 1964, Chapter 7.

References

[3] C. T. Sah, R. N. Noyce, and W. Shockley, "Carrier Generation and Recombination in *p-n* Junctions and *p-n* Junction Characteristics," *Proc. IRE*, **45**, 1228–1243 (Septemper 1957).

[4] C. T. Sah, "Effect of Surface Recombination and Channel on *p-n* Junction and Transistor Characteristics," *IRE Trans. Electron Devices*, **9**, 94–108 (January 1962).

[5] C. T. Sah, "Effect of Electrons and Holes on the Transition Layer Characteristics of Linearly Graded *p-n* Junctions," *Proc. IRE*, **49**, 603–618 (March 1961).

[6] J. J. Ebers and J. L. Moll, "Large Signal Behavior of Junction Transistors," *Proc. IRE*, **42**, 1761–1772 (December 1954).

[7] C. T. Sah, "A New Semiconductor Tetrode—The Surface Potential Controlled Transistor," *Proc. IRE*, **49**, 1623–1634, (November 1961).

[8] P. J. Coppen and W. T. Matzen, "Distribution of Recombination Current in Emitter-Base Junctions of Silicon Transistors," *IRE Trans. Electron Devices*, **9**, 75–81 (January 1962).

[9] J. E. Iwerson et al., "Low Current Alpha in Silicon Transistors," *IRE Trans. Electron. Devices*, **9**, 474–478 (November 1962).

[10] V. G. K. Reddi and C. Bittman, "Micropower Functional Electronic Blocks," Interim Technical Reports 1, 2, 3, 4, Contract No. AF33 (615)-3010, August 1965, January, March, May 1966.

[11] K. D. Kang, "Detailed Study of Deleterious Effects on Silicon Transistors," Fourth Technical Report, Contract No. AF30 (602)-3244 October 1964.

[12] R. H. Beeson, "A Complete Transistor Equivalent Circuit," *Proc. IRE* (Correspondence), **49**, 825–826 (April 1961).

[13] J. L. Moll, "Large-Signal Transient Response of Junction Transistors," *Proc. IRE*, **42**, 1773–1784 (December 1954).

[14] D. J. Hamilton et al., "Comparison of Large Signal Models for Junction Transistors," *IEEE*, **52**, 239–248 (March 1964).

[15] R. Beaufoy and J. J. Sparkes, "The Junction Transistor as a Charge Controlled Device," *ATEJ*, **13**, 310–327 (October 1957).

[16] J. G. Linvill and J. F. Gibbons, "Transistors and Active Circuits," McGraw-Hill, New York, 1961. Also private communication from Dr. J. G. Linvill.

[17] J. D. Meindl et al., "Static and Dynamic Performance of Micropower Transistor Logic Circuits," *Proc. IEEE*, **52**, 1575–1580 (December 1964).

[18] C. D. Simmons, "High Speed Micro-energy Switching," *Solid State J.*, (September–October 1960).

[19] S. Steiner, "Turn-off Transition Mechanism," *Solid State Design*, pp. 29–34 (June 1962).

[20] J. D. Meindl et al., "Static and Dynamic Performance of Micropower Transistor Linear Amplifiers," in *Micropower Electronics*, E. Keonjian, Ed., Macmillan, New York, 1964.

[21] P. E. Gray et al., *SEEC Notes 2: Physical Electronics and Circuit Models of Transistors*, Wiley, New York, 1965, Chapter 8.

[22] R. L. Pritchard et al., "Transistor Internal Parameters for Small-Signal Representation," *Proc. IRE*, **49**, 725–738 (April 1961).

[23] J. L. Moll, "Relations Between Minimum Required Power Density and Frequency Response for Present and Future Semiconductor Triode Amplifiers," in *Micropower Electronics*, E. Keojian, Ed., Macmillan, New York, 1964.

[24] W. W. Gaertner, "Nanowatt Devices," *Proc. IEEE*, **53**, 592–604 (June 1965).

[25] J. D. Meindl and P. H. Hudson, "Low Power Linear Circuits," *IEEE J. Solid-State Circuits*, **1**, 100–111 (December 1966).

[26] J. M. Goldey and R. Ryder, "Are Transistors Approaching Their Maximum Capabilities?", Digest of Technical Papers, 1963 International Solid-State Circuits Conference, p. 20 (February 1963).
[27] W. Shockley, "A Unipolar Field Effect Transistor," *Proc. IRE*, **40**, 1365–1376 (November 1952).
[28] G. C. Dacey and I. M. Ross, "Unipolar Field Effect Transistor," *Proc. IRE*, **41**, 970–979 (August 1953).
[29] R. R. Bockemuehl, "Analysis of Field Effect Transistors with Arbitrary Charge Distribution," *IEEE Trans. Electron. Devices*, **10**, 31–34 (January 1963).
[30] R. D. Middlebrook, "A Simple Derivation of FET Characteristics," *Proc. IEEE* (Correspondence), **51**, 1146–1147 (August 1963).
[31] L. J. Sevin, *Field-Effect Transistors*, McGraw-Hill, New York, 1965, Chapter 1.
[32] I. Richer, "Input Capacitance of Field-Effect Transistors," *Proc. IEEE* (Correspondence), **51**, 1249–1250 (September 1963).
[33] J. R. Hauser, "Small Signal Properties of Field Effect Devices," *IEEE Trans. Electron. Devices*, **12**, 605–618 (December 1965).
[34] H. K. J. Ihantola, "Design Theory of a Surface Field-Effect Transistor," Stanford Electronics Lab, Stanford University, Stanford, California, Technical Report. 1661-1, Contract No. AF 33(616)-7726 (September 1961).
[35] S. R. Hofstein and F. P. Heiman, "The Silicon Insulated-Gate Field-Effect Transistor," *Proc. IEEE*, **51**, 1190–1202 (September 1963).
[36] C. T. Sah, "Characteristics of the Metal-Oxide-Semiconductor, Transistors," *IEEE Trans. Electron. Devices*, **11**, 324–345 (July 1964).
[37] C. T. Sah and H. C. Pao, "The Effects of Fixed Bulk Charge on the Characteristics of Metal-Oxide-Semiconductors Transistors," *IEEE Trans. Electron. Devices*, **13**, 393–409 (April 1966).
[38] P. Gardner et al., "Application of Tunnel Diodes to Micropower Logic Circuits," Correspondence, *Solid-State Electronics*, Pergamon Press, G. B., **7**, 547–551 (1964).
[39] K. Tarnay, "Approximation of Tunnel-Diode Characteristics," *Proc. IRE* (Correspondence), **50**, 202–203 (February 1962).
[40] W. F. Chow, *Principles of Tunnel Diode Circuits*, Wiley, New York, 1964.
[41] A. Ferenderi and W. H. Ko, "A Two-Term Analytical Approximation of Tunnel-Diode Static Characteristics, *Proc. IRE* (Correspondence), **50**, 1852–1853 (August 1962).
[42] C. F. Parks, "Resistors for Micropower Circuits," Contract No. DA 28-043-AMC-01524(E) (November 1965, March, July 1966).
[43] R. P. Donovan et al., "Integrated Silicon Device Technology, Vol. I—Resistance," Contract No. AF 33(657)-10340 (June 1963).
[44] F. M. Warner et al., *Integrated Circuits—Design Principles and Fabrication*, McGraw-Hill, New York, 1965, Chapter 10.
[45] *Ibid.*, Chapter 13.
[46] G. E. Moore, "Semiconductor Integrated Circuits," in *Microelectronics*, E. Keonjian, Ed., McGraw-Hill, New York, 1963.
[47] L. Vadasz, "The Use of MOS Structure for the Design of High Value Resistors in Monolithic Integrated Circuits," *IEEE Trans. Electron. Devices*, **13**, 459–465 (May 1966).
[48] L. K. Monteith et al., "Integrated Silicon Device Technology, Vol. II—Capacitance," Contract No. AF 33(657)-10340 (October 1963).

Chapter 2

The Quiescent Point

The quiescent point or dc operating point of a micropower circuit should be rather carefully determined to permit the minimum dc power consumption for a specified ac performance. The nominal dc operating point of a micropower circuit can be drastically altered by such factors as (a) initial tolerances in the values of the parameters of both active and passive circuit elements, (b) the temperature variations of all circuit elements, (c) changes in supply voltages, (d) radiation environments, and (e) other deleterious environments as well as aging. This chapter is concerned with the problems of establishing and maintaining a desired quiescent point in a micropower circuit while purposely limiting the bias network power dissipation. It is assumed that the micropower circuits are intended for use in processing ac signals and are not to be used as direct current amplifiers.

2.1 DEVICE PARAMETERS

A transistor model derived from (1.10) and (1.16) that applies in the active region is illustrated in Figure 2.1. Changes in the values of the parameters I_{CBO}, V_{BE}, and h_{FE} tend to alter the quiescent point (I_C, V_{CE}) of a circuit. Since the influence of excess junction current is more pronounced in the micropower range, initial tolerances, temperature variations, and other changes in V_{BE} and h_{FE} are generally less predictable than in the normal low-power range. I_{CBO} is a larger fraction of the operating currents, I_B and I_C, whereas V_{BE} is often a larger fraction of the supply voltage V_{CC}. Such

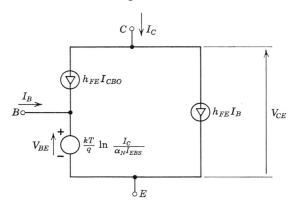

Figure 2.1 Static transistor model in active region.

factors serve to intensify the quiescent point stability problem in a micropower circuit. In addition, initial tolerances, temperature variations, and other changes in the nominal values of the resistors in the circuit must be considered in its dc design.

2.2 SPECIFIED POWER DESIGN

The problem of stabilizing the dc operating point of low-power transistor circuits has been investigated by a number of authors [1–10]. In essence, the dc design of a micropower circuit does not differ from that of a typical low-power circuit; however, the primary emphasis placed on minimum power consumption does evoke certain unusual features [11].

Emitter Degeneration

The schematic diagram in Figure 2.2 represents a widely used low-power circuit configuration based on the "discrete part" technologies. Emitter degeneration is used to stabilize the quiescent point. The Kirchhoff voltage equations that describe the circuit are

$$V_{CC} = I_C R_C + V_{CE} + I_E R_E \qquad (2.1a)$$

$$V_{CC} = I_2 R_2 + (I_2 - I_B) R_1 \qquad (2.1b)$$

$$0 = V_{BE} + I_E R_E - (I_2 - I_B) R_1. \qquad (2.1c)$$

(The assumed positive direction of I_E is out of the emitter in this instance.) In a typical design V_{CC} is fixed by available battery voltages whereas I_C,

R_C, and a minimum value of V_{CE} are determined by ac performance requirements. Some flexibility exists in the selection of R_E by trading between V_{CE} and $I_E R_E$ in (2.1a). By selecting I_2, which specifies the nominal quiescent power, the design can be completed since (2.1b) and (2.1c) readily yield R_1 and R_2. Typically, selecting I_2 so that $5 I_B \leq I_2 \leq I_C/5$ provides a useful design whose base-circuit power dissipation is insignificant.

With the nominal values of the circuit elements established by the preceding "specified power design," (2.1) may be solved to give

$$I_C = \frac{\dfrac{R_1}{R_1 + R_2} V_{CC} - V_{BE} + I_{CBO}\left(\dfrac{R_1 R_2}{R_1 + R_2} + R_E\right)}{\dfrac{1}{h_{FE}}\left(\dfrac{R_1 R_2}{R_1 + R_2} + R_E\right) + R_E}. \qquad (2.2)$$

This equation can be used to compute the resultant value of I_C for any set of initial tolerances, temperature variations, and other changes from the nominal values of I_{CBO}, V_{BE}, h_{FE}, R_C, R_E, R_1, R_2, and V_{CC} by simply substituting the actual values of these quantities into (2.2). By computing the worst case [10,11] values of I_C at the lower and upper operating temperature limits of the circuit, the adequacy of the stabilization network design can be determined. If the change in I_C is too large, an increase in R_E or I_2 or both often leads to improved quiescent point stability. The value of V_{CE} corresponding to a particular I_C can be computed from (2.1a).

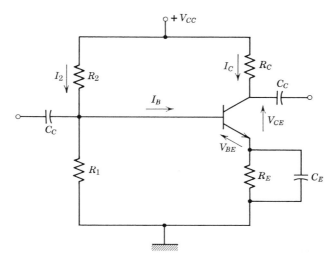

Figure 2.2 Single stage circuit with emitter degeneration.

Emitter and Collector Degeneration

A second circuit configuration often used for discrete part designs is illustrated in Figure 2.3. Both emitter and collector degeneration are used to achieve a stable quiescent point. The nominal operating point can be established following the procedure used for the circuit in Figure 2.2. The

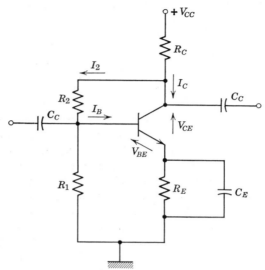

Figure 2.3 Single stage circuit with emitter and collector degeneration.

stability can be determined by computing the worst case values of I_C from

$$I_C \simeq \frac{\dfrac{R_1}{R_1 + R_2} V_{CC} - V_{BE} + I_{CBO}\left(\dfrac{R_1 R_2}{R_1 + R_2}\right)}{\dfrac{1}{h_{FE}} \dfrac{R_1 R_2}{R_1 + R_2}\left[1 + h_{FE} \dfrac{R_C + R_E}{R_2}\right] + R_E}, \qquad (2.3)$$

which assumes that $I_E \simeq I_C \gg I_2$.

Collector Degeneration and Diode Compensation

The circuit configuration [12,13] illustrated in Figure 2.4 offers three salient advantages for the silicon monolithic integrated circuit technology:

1. The absence of an emitter feedback resistor obviates the need for a large bypass capacitor such as C_E in Figure 2.3.

Specified Power Design

2. The total resistance in the circuit is often small in comparison to that required in the circuits in Figure 2.2 or 2.3 for equivalent overall performance. Particularly in micropower circuits, this alleviates the need for large substrate areas for resistor fabrication.

3. The strong thermal coupling and excellent matching between the diode-connected transistor and the active transistor permit excellent V_{BE} compensation of the active transistor.

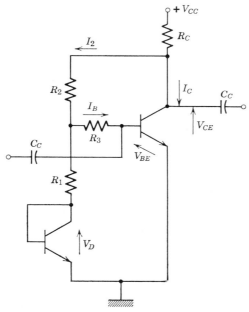

Figure 2.4 Single stage circuit with collector degeneration and diode compensation.

When we establish the nominal quiescent point, R_3 and I_2 can be selected and the corresponding values of R_1 and R_2 computed from the Kirchhoff voltage equations of the circuit. Both collector degeneration and diode compensation are used to stabilize the quiescent point. Assuming $I_E \simeq I_C$,

$$I_C \simeq \frac{\dfrac{R_1}{R_1 + R_2 + R_C}(V_{CC} - V_D) - (V_{BE} - V_D) + I_{CBO}\left[R_3 + \dfrac{R_1(R_2 + R_C)}{R_1 + R_2 + R_C}\right]}{\dfrac{R_1 R_C}{R_1 + R_2 + R_C} + \dfrac{1}{h_{FE}}\left[R_3 + \dfrac{R_1(R_2 + R_C)}{R_1 + R_2 + R_C}\right]}.$$

(2.4)

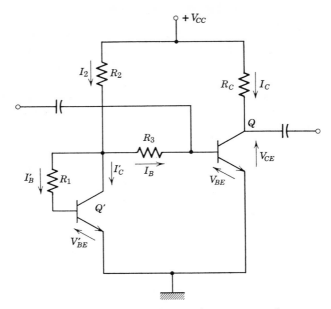

Figure 2.5 Single stage with transistor compensation.

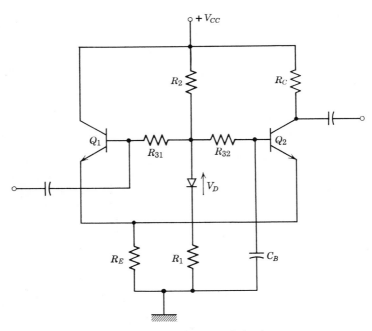

Figure 2.6 Emitter coupled pair.

Transistor Compensation

A very useful circuit configuration [14] for monolithic integrated circuits is illustrated in Figure 2.5. The advantages are similar to those discussed in the preceding subsection. Excellent matching and temperature tracking of V_{BE} and h_{FE} between the compensating and active transistors permit a stable quiescent current without the use of feedback for stabilization.

$$I_C \simeq \frac{\dfrac{R_1}{R_1 + (h'_{FE} + 1)R_2}[V_{CC} - V'_{BE} - h'_{FE}I'_{CBO}R_2] - (V_{BE} - V'_{BE}) + I_{CBO}\left[R_3 + \dfrac{R_1 R_2}{R_1 + (h'_{FE} + 1)R_2}\right]}{\dfrac{1}{h_{FE}}\left[R_3 + \dfrac{R_1 R_2}{R_1 + (h'_{FE} + 1)R_2}\right]}$$

$$\simeq \frac{V_{CC} - V'_{BE}}{R_2}. \tag{2.5}$$

Emitter Coupled Pair

The emitter coupled pair of Figure 2.6 is a circuit configuration that is amenable to monolithic integrated circuit technology and that achieves a high degree of quiescent point stability through close matching and thermal coupling of the circuit elements as well as substantial degeneration from R_E. Assuming an excellent match of similar elements, $h_{FE1} = h_{FE2} = h_{FE}$, $V_{BE1} = V_{BE2} = V_{BE}$, $I_{CBO1} = I_{CBO2} = I_{CBO}$, and $R_{31} = R_{32} = R_3$ so that $I_{C1} = I_{C2} = I_C$, for which

$$I_C \simeq \frac{\dfrac{R_1}{R_1 + R_2}(V_{CC} - V_D) - (V_{BE} - V_D) + I_{CBO}\left(2\dfrac{R_1 R_2}{R_1 + R_2} + R_3\right)}{\dfrac{1}{h_{FE}}\left(2\dfrac{R_1 R_2}{R_1 + R_2} + R_3\right) + 2R_E} \tag{2.6}$$

provides the basis for determining operating point stability. A small base-circuit bypass capacitor, C_B, is used in this circuit. However, the dc coupling of the emitters avoids the need for a coupling capacitor.

Feedback Pair

The feedback pair in Figure 2.7 provides a simple bias network suitable for both discrete part and monolithic fabrication techniques.

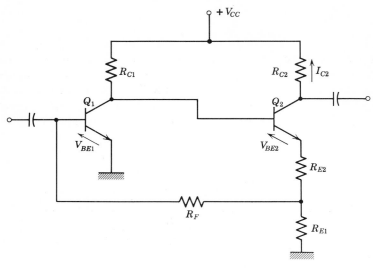

Figure 2.7 Feedback pair.

Assuming $I_E \simeq I_C \gg I_B$,

$$I_{C2} \simeq \frac{V_{CC} - V_{BE2} + h_{FE1}R_{C1}[V_{BE1}/R_E - I_{CBO1}]}{(R_{E1} + R_{E2}) + h_{FE1}(R_{E1}R_{C1}/R_F)} \qquad (2.7)$$

provides a basis for evaluation of quiescent point stability. The dc coupling between Q_1 and Q_2 avoids the need for a coupling capacitor.

Multistage Direct Coupled Circuits

Although solution of the simultaneous equations can become somewhat tedious, the specified power design procedure can be applied to a direct coupled circuit of any level of complexity by following the pattern discussed in this section. It is a straightforward procedure for handling the problem of establishing and maintaining the desired operating points for multistage direct coupled amplifiers of interest in the micropower range.

2.3 SPECIFIED STABILITY DESIGN

In the previous section the nominal quiescent power of a micropower circuit was specified as a convenient means of arriving at a design whose stability was then evaluated and, if necessary, adjusted by repeated design trials. An alternative approach to the stability problem [10, 11] is explicitly to specify the limiting operating points and then calculate the circuit design required to achieve the specified stability. In certain instances—for example, in the design of an ac amplifier whose gain and terminal impedances are temperature compensated [15]—explicit definition of the limiting quiescent points proves helpful.

The specified stability design can be illustrated with a simple example. From (2.1),

$$V_B = V_{BE} + I_C R_E + (R_B + R_E)I_B, \qquad (2.8)$$

where

$$V_B = \frac{R_1}{R_1 + R_2} V_{CC} \quad \text{and} \quad R_B = \frac{R_1 R_2}{R_1 + R_2}. \qquad (2.9)$$

If V_B, R_B, and R_E are insensitive to temperature, then (2.8) gives

$$V_B = V_{BEx} + I_{Cx} R_E + (R_B + R_E)I_{Bx} \qquad (2.10a)$$

and

$$V_B = V_{BEy} + I_{Cy} R_E + (R_B + R_E)I_{By}, \qquad (2.10b)$$

where the subscript x denotes the upper operating temperature limit of the circuit T_x and y denotes the lower limit T_y. From (2.10)

$$R_B = \frac{(V_{BEx} + I_{Ex} R_E) - (V_{BEy} + I_{Ey} R_E)}{I_{By} - I_{Bx}} \qquad (2.11a)$$

and

$$V_B = \frac{(V_{BEx} + I_{Ex} R_E)I_{By} - (V_{BEy} + I_{Ey} R_E)I_{Bx}}{I_{By} - I_{Bx}}, \qquad (2.11b)$$

which in conjunction with (2.9) define the values of R_1 and R_2, for a selected value of R_E, for which the quiescent current is I_{Cx} at T_x and I_{Cy} at T_y.

The effects of initial tolerances in the resistors can be included in the specified stability analysis by expressing the actual value of a nominal resistance R as $R(1 \pm \delta_R)$ where δ_R is the percent of tolerance expressed in decimal form. Temperature variations can be included by using $\gamma_R = R_x/R_y$ where $R_x = R @ T_x$ and $R_y = R @ T_y$.

The specified stability design procedure can be used for any of the circuits of Section 2.3. A computer-aided design is desirable when using this procedure.

2.4 EFFECTS OF IMPERFECT ELEMENT ISOLATION

The schematic diagram in Figure 2.8 illustrates the parasitic currents associated with the isolation junctions of silicon monolithic integrated

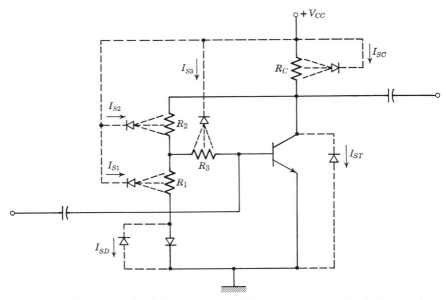

Figure 2.8 Single stage circuit illustrating parasitic reverse currents of isolation junctions.

circuits. At high temperatures and in radiation environments these parasitic reverse junction currents can influence the operating point stability of a micropower circuit. The parasitic currents associated with high-value diffused silicon resistors are especially important in view of the usually large isolation junction areas of these resistors. The static parasitic currents associated with monolithic capacitors can also become important. In this analysis, however, external discrete part capacitors with negligible leakage current are assumed.

The Kirchhoff equations of the circuit can be formulated on the basis of the model of Figure 2.9. An approximate solution to these equations yields the result of (2.4) with the following additional terms in the numerator:

$$I_{S3}\left[\frac{R_3}{2} + \frac{R_1(R_2 + R_C)}{R_4}\right] + I_{S2}\left[\frac{R_1(R_C + R_2/2)}{R_4}\right]$$

$$+ I_{S1}\left(\frac{R_1}{2}\frac{R_2 + R_C}{R_4}\right) + I_{SC}\frac{R_1 R_C/2}{R_4} - I_{ST}\frac{R_1 R_C}{R_4} \quad (2.12)$$

where $R_4 = R_1 + R_2 + R_C$. For small geometry transistors and large value diffused resistors, it is entirely possible that the dominant source of quiescent point drift will be resistor isolation junction reverse currents as reflected in (2.12).

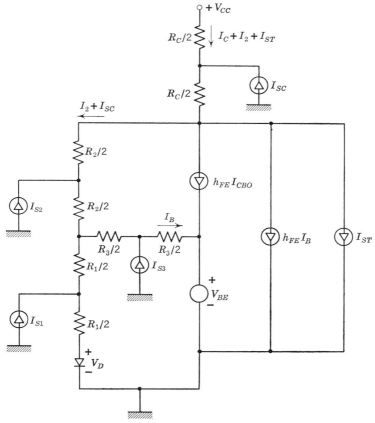

Figure 2.9 DC model for circuit of Figure 2.8.

REFERENCES

[1] R. F. Shea, "Transistor Operation: Stabilization of Operating Points," *Proc. IRE*, **40**, 1435–1437 (November 1952).
[2] R. F. Shea et al., *Principles of Transistor Circuits*, Wiley, New York, 1953.
[3] R. F. Shea, *Transistor Circuits Engineering*, Wiley, New York, 1957.
[4] S. K. Ghandhi, "Analysis and Design of Transistor Bias Networks," *Proc. Natl. Electron. Conf.*, **12**, 491–504 (1956).
[5] A. W. Lo et al., *Transistor Electronics*, Prentice-Hall, Engelwood Cliffs, N.J., 1955.
[6] H. C. Lin and A. A. Barco, "Temperature Effects in Circuits Using Junction Transistors," *Transistors I*, pp. 369–402, RCA Laboratories, Princeton, N.J. (1956).
[7] L. M. Vallese, "Temperature Stabilization of Transistor Amplifiers," *IEEE Trans. Commun. Electron.* No. 26, 379–384 (September 1956).

[8] S. Schwartz, *Selected Semiconductor Circuits Handbook*, Wiley, New York, 1960.
[9] C. L. Searle et al., "Elementary Circuit Properties of Transistors," *SEEC*, **3**, Wiley, New York, 1964, Chapter 5.
[10] J. D. Meindl and O. Pitzalis, "Optimum Stabilization Networks for Functional Electronic Blocks," *Proc. NEC*, **16**, 576–590 (October 1960).
[11] J. D. Meindl et al., "Static and Dynamic Performance of Micropower Transistor Linear Amplifiers," *Micropower Electronics*, E. Keonjian ed., Macmillan, New York, 1964.
[12] J. Feit and R. McGinnis, "Small signal functional circuits," USAECOM, Fort Monmouth, N.J., Contract No. DA 28-043 AMC-00150(E), Final Report, July 1964–June 1965.
[13] H. Fischler, "Noise Improvements in Transistors," U.S. Patent 3,241,083, 15 March 1966.
[14] R. J. Widlar, "Some Circuit Design Techniques for Linear Integrated Circuits," *IEEE Trans. Circuit Theory*, **12**, 586–590 (December 1965).
[15] J. D. Meindl and O. Pitzalis, "AC Temperature Compensation and Integrated Amplifiers," *IEEE Trans. Commun. Electron.*, No. **83**, 579–584 (November 1964).

Chapter 3

Low–Frequency Amplifiers

The principal question of interest in this chapter is "What are the primary design constraints that limit the minimum quiescent power of a low-frequency amplifier?" A low-frequency amplifier is identified as a circuit whose active devices can be represented adequately by models that are essentially free of reactive elements. Figure 3.1 illustrates the block diagram of a multistage low-frequency amplifier [1–4]. The criteria that determine the minimum quiescent power of a particular stage depend upon its location. For example, the primary design constraint may be input impedance for the first stage, power gain for the intermediate stages, and power output for the driver stage and output stage. Often, because of the relatively high operating power requirements of the driver and output stages, minimum power designs for input and intermediate stages consist largely of restricting their quiescent

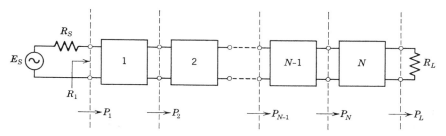

Figure 3.1 Block diagram of multistage low-frequency amplifier.

power to negligible values compared to the requirements of the driver and output stages.

3.1 INPUT STAGES

The basic design constraints for an input stage of a low-frequency amplifier are input impedance, gain, and noise figure. In most instances either the required input impedance or noise figure exerts the primary influence on the quiescent power. The effect of input impedance is discussed in this section, and noise figure in Chapter 6.

Impedance Matching

In certain instances it may be desirable to match the internal impedance of a signal source, R_s, to the input impedance of an amplifier, R_1, in order to achieve maximum power transfer from the source to the amplifier. Such a condition yields maximum transducer power gain (Figure 3.1)

$$G_T = \frac{P_L}{P_{AVS}} = \frac{P_1}{P_{AVS}} \frac{P_2}{P_1} \cdots \frac{P_N}{P_{N-1}} \frac{P_L}{P_N}$$

$$= \frac{P_1}{P_{AVS}} G_1 G_2 \cdots G_{N-1} G_N \qquad (3.1)$$

for a constant (actual) power gain $G_1, G_2, \ldots, G_{N-1}, G_N$ for the amplifier, since the ratio of the input power $P_1 = R_1(E_S/R_1 + R_s)^2$ to the available power of the source $P_{AVS} = E_S^2/4R_S$ reaches a maximum value, $P_1/P_{AVS} = 1$, for $R_S = R_1$.

For a micropower common-emitter input stage [5–9], as illustrated in Figure 3.2, the condition of maximum power transfer from source to amplifier requires

$$R_S = R_1 = r_{be} = h_{fe0} \frac{kT}{qI_C} \qquad (3.2)$$

or an input stage collector current

$$I_C = h_{fe0} \frac{kT}{q} \frac{1}{R_S}. \qquad (3.3)$$

From (3.3) it is evident that the quiescent current (or power) of the common-emitter input stage is inversely proportional to R_S. For small values of R_S, the I_C required for matching may be tolerated as long as it does not compare

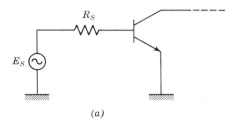

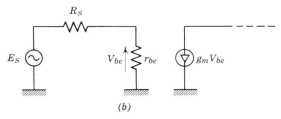

Figure 3.2 Input stage: (a) ac schematic diagram; (b) ac equivalent circuit.

appreciably with the total current drain of the amplifier. When the required value of I_C becomes too large compared to total current drain, it may well be advantageous to accept a poor impedance match at the input transistor and to compensate for it by increasing the gain of the amplifier in order to maintain the required overall performance at an acceptable power drain. Also, ac shunt feedback can be used to reduce input impedance as suggested by Figure 2.3.

Minimum Loading

In certain instances it may be desirable to design an input stage to present a very high impedance to the source, usually to minimize the loading of the source by the amplifier. As the required value for R_1 increases, (3.2) indicates that substantial values (for example, $r_{be} = 2.6 \times 10^6 \, \Omega$ for $h_{fe0} = 100$ at $I_C = 1.0 \, \mu A$) can be achieved using readily available micropower transistors. Still larger values of R_1 can be obtained by further reducing collector current and introducing degenerative ac emitter feedback as illustrated in Figure 3.3. The input impedance of the circuit of Figure 3.3 is

$$R_1 \simeq r_{be} + (h_{fe0} + 1)R_e$$

$$\simeq h_{fe0} R_e, \qquad R_e \gg \frac{1}{g_m} = \frac{kT}{qI_C} \tag{3.4}$$

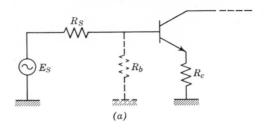

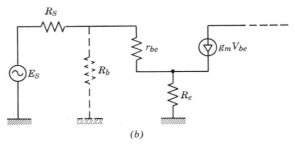

Figure 3.3 Input stage with ac emitter degeneration: (a) ac schematic; (b) ac equivalent circuit.

(so that at $I_C = 1.0\ \mu A$ with $1/g_m = 26 \times 10^3\ \Omega$ and $R_e = 10/g_m = 260 \times 10^3$, $R_1 \simeq 26 \times 10^6\ \Omega$) for $R_b \gg h_{fe0}R_e$. As R_1 is increased by enlarging R_e, the shunting effect of the equivalent ac resistance of the base biasing resistors, R_b, indicated in Figure 3.3 may become significant. In this instance the input admittance is given by

$$\frac{1}{R_1} \simeq \frac{1}{R_b} + \frac{1}{r_{be} + h_{fe0}R_e}$$

$$\simeq \frac{1}{R_b} + \frac{1}{r_{be}(1 + g_m R_e)}. \tag{3.5}$$

The shunting effect of R_b can be reduced by the use of a feedback technique called bootstrapping [10,11] illustrated in Figure 3.4. The input admittance of this circuit is

$$\frac{1}{R_1} \simeq \frac{1}{R_b(1 + g_m R_e)} + \frac{1}{r_{be}(1 + g_m R_e)} \tag{3.6}$$

for $R_b \simeq R_{B3} \gg R_{B1} \parallel R_{B2} \gg R_e$ and $R_1 \ll r_c$, the transistor collector-to-base resistance (see Figure 1.6). The admittance given by (3.6) can be reduced

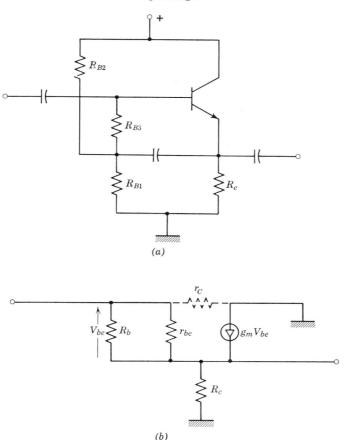

Figure 3.4 Common collector input stage with "bootstrapped" bias network.

until r_c determines the limiting value of the input admittance for the circuit in Figure 3.4, after which r_c can be bootstrapped [10,11].

FET Inputs

The junction gate field effect transistor (JGFET) provides a simple and effective means for achieving high-impedance input stages for low-frequency micropower amplifiers [12,13] as illustrated in Figure 3.5. The input impedance of the circuit is determined by the gate-return resistor R_b. Bootstrapping techniques can be used following the pattern illustrated in Figure 3.4 to increase R_b effectively.

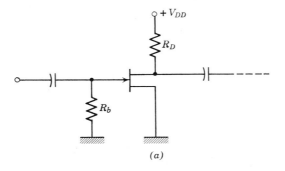

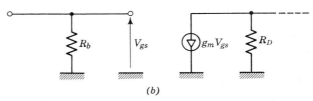

(b)

Figure 3.5 Junction gate FET input stage.

3.2 INTERMEDIATE STAGES

In this section it is assumed that the primary design constraint of the intermediate stages of a low-frequency amplifier is overall amplification, A_{vo}^*.

Specified Number of Stages

For a specified number of iterative intermediate stages [2,3], n, the voltage and current amplification per stage are

$$A_{v0} = A_{i0} = (A_{v0}^*)^{1/n} = (A_{i0}^*)^{1/n} = -\frac{v_n}{v_{n-1}}. \tag{3.7}$$

From Figure 3.6, the amplification per stage can be expressed approximately as

$$A_{v0} = h_{fe0}\frac{R_C}{R_C + r_{be}}$$

$$\simeq g_m R_C \quad \text{for} \quad r_{be} \gg R_C \tag{3.8}$$

or

$$\simeq h_{fe0} \quad \text{for} \quad r_{be} \ll R_C.$$

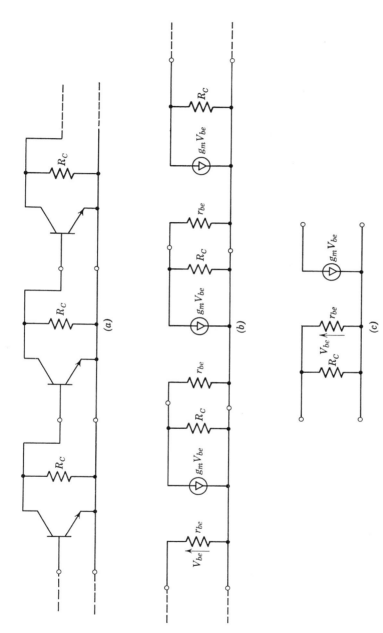

Figure 3.6 Low-frequency iterative amplifier: (*a*) ac schematic diagram; (*b*) multistage equivalent circuit; (*c*) iterative stage equivalent circuit.

For a bias network configuration similar to that of Figures 2.4 and 2.5, the dc load equation is

$$V_{CC} \simeq I_C R_C + V_{CE}, \qquad (3.9)$$

where V_{CC} is fixed. The total quiescent power drain of n iterative stages is

$$P_{DC}^* \simeq n V_{CC} I_C, \qquad (3.10)$$

assuming that the base bias network power drain is negligible. For a prescribed quiescent power P_{DC}^*, which often can be negligible compared to the power requirements of the driver and output stages, (3.10) yields the transistor operating current I_C; (3.8) yields the collector load resistance R_C; and (3.9) then yields the collector voltage V_{CE} of the iterative intermediate stage.

Specified Total Resistance

From (3.9) and (3.10) it is evident that as the prescribed value of P_{DC}^* is reduced, R_C tends to increase proportionately in order to fulfill the requirement $V_{BE} \leq V_{CE} \leq V_{CC}$. Especially in micropower monolithic integrated circuits, the large resistance values (and consequently the large substrate areas) required as P_{DC}^* is reduced can present rather substantial technological and economic problems. In such instances it may be desirable to limit the total resistance of the circuit to a prescribed value.

For example, in the circuit configuration in Figure 3.7 the amplification per stage is

$$A_{v0} = \frac{g_m}{(G_C + G_B) + g_m/h_{fe0}} \qquad (3.11)$$

and the total quiescent power for $I_C \simeq I_{CB}$ is

$$P_{DC}^* \simeq 2n V_{CC} I_C$$

$$\simeq 2n V_{CC} \left[\gamma (G_C + G_B) \frac{A_{v0}}{1 - A_{v0}/h_{fe0}} \right] \qquad (3.12)$$

after substitution of (3.11). The total circuit resistance is

$$R_T^* \simeq n R_T \simeq 2n(R_C + R_B), \qquad (3.13)$$

assuming $R_C \simeq R_{CB}$.

Combining (3.13) with (3.12) yields

$$P_{DC}^* = 4\gamma V_{CC} \frac{1}{R_T^*} (2 + \theta + 1/\theta) n^2 \frac{A_{v0}}{1 - A_{v0}/h_{fe0}} \qquad (3.14a)$$

or

$$P_{DC}' = \frac{P_{DC}^*}{4\gamma V_{CC} \dfrac{1}{R_T^*} (2 + \theta + 1/\theta)} = n^2 \frac{(A_{v0}^*)^{1/n}}{1 - (A_{v0}^*)^{1/n}/h_{fe0}}, \qquad (3.14b)$$

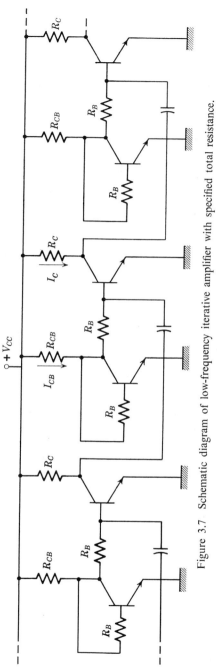

Figure 3.7 Schematic diagram of low-frequency iterative amplifier with specified total resistance.

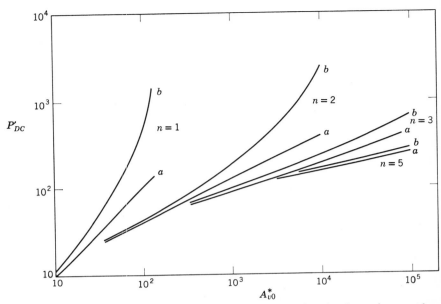

Figure 3.8 Normalized quiescent power versus overall amplification for various numbers of stages for the amplifier configuration of Figure 3.7.

where $\theta = R_B/R_C$. Figure 3.8 displays a normalized plot of (3.14b) for $h_{fe0} = 120$. The a curves reflect the quiescent power for $A_{v0} \ll h_{fe0}$; the b curves reflect P_{DC}^* for typical designs where $A_{v0} \rightarrow h_{fe0}$. It is evident from Figure 3.8 that $A_{v0} \ll h_{fe0}$ is a desirable condition for small P_{DC}^*. In addition, by increasing the number of stages n for a given A_{v0}^*, substantial reductions in P_{DC}^* are possible [14]. For example, at $A_{v0}^* = A_{v0}{}^n = 10^2$ the b curves indicate the normalized quiescent power $P'_{DC} = 600$ for $n = 1$ can be reduced to $P'_{DC} \simeq 44$ for $n = 2$.

From (3.14), $\partial P_{DC}^*/\partial \theta$ indicates that P_{DC}^* is minimum for $\theta = 1$ or $R_B = R_C$.

For $A_{v0} \ll h_{fe0}$, $\partial P_{DC}^*/\partial n$ indicates

$$A_{v0} \simeq \epsilon^2 = 7.40 = 17.4 \text{ db} \quad \text{and} \quad n \simeq \tfrac{1}{2} \ln A_{v0}^* \qquad (3.15)$$

are the optimum amplification per stage and number of stages, respectively, for minimum P_{DC}^* for a specified total resistance.

3.3 DRIVER STAGES

A simplified schematic diagram of a capacitor-coupled class A driver or output stage is illustrated in Figure 3.9. Among the more common

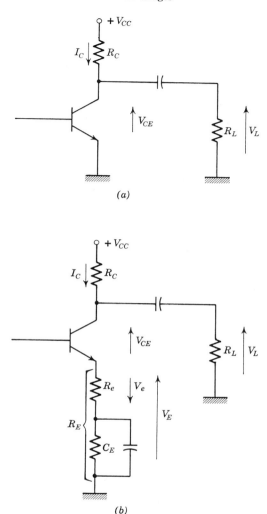

Figure 3.9 Driver stage schematic diagram (a) without and (b) with ac and dc emitter feedback.

constraints that must be observed in the design of this circuit are (a) a fixed supply voltage, V_{CC}; (b) a specified load resistance, R_L; (c) a specified peak ac output voltage, V_L; (d) a restricted amount of harmonic distortion; and (e) a prescribed operating temperature range. The problem of particular interest in this section is how to satisfy these constraints while minimizing the quiescent power, P_{DC}, required by the circuit.

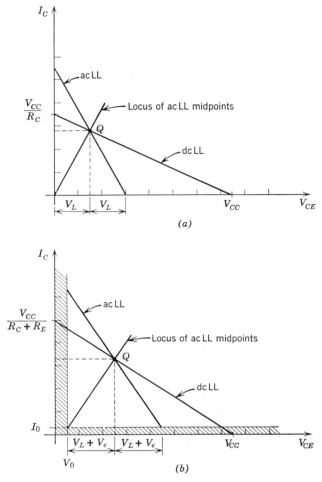

Figure 3.10 Class A loadline diagrams (a) without and (b) with ac and dc emitter feedback

Basic Design [1,15]

The dc load line (dcLL) of the micropower circuit of Figure 3.9a, illustrated in Figure 3.10a, is described by

$$V_{CC} = I_C R_C + V_{CE}. \tag{3.16}$$

In order to provide the specified V_L for minimum P_{DC}, the circuit must be designed so that the dc operating point Q bisects the ac load line (acLL)

given by
$$I_C = -(G_C + G_L)V_{CE}, \qquad (3.17)$$
which represents the acLL in the (I_C, V_{CE}) plane. The locus of all acLL midpoints is given by
$$I_C = (G_C + G_L)V_{CE}. \qquad (3.18)$$
Neglecting for the moment the effects of cutoff and saturation in the transistor, we note from Figure 3.10a that $V_{CE} = V_L$ permits maximum efficiency so that (3.16) and (3.18) yield
$$I_C = \frac{V_{CC} - V_L}{V_{CC} - 2V_L} V_L G_L \qquad (3.19)$$
and
$$R_C = \frac{V_{CC} - 2V_L}{V_L G_L}, \qquad (3.20)$$
the smallest I_C and largest R_C, respectively, which satisfy the design constraints on V_{CC}, R_L, and V_L.

The efficiency of the circuit of Figure 3.9a can be expressed as
$$\eta \simeq \frac{P_L}{P_{DC}} = \frac{\tfrac{1}{2}V_L^2 G_L}{V_{CC} I_C} = \frac{V_L(V_{CC} - 2V_L)}{2V_{CC}(V_{CC} - V_L)} \qquad (3.21)$$
after substitution of (3.19). From (3.21), $\partial \eta / \partial V_{CC}$ indicates that for $V_{CC} = (2 + \sqrt{2})V_L$—and therefore $I_C = (1 + 1/\sqrt{2})V_L G_L$ and $R_C = \sqrt{2}R_L$—η reaches its maximum value of 8.59%, which can be achieved only if V_{CC} is adjusted to its optimum value for the specified V_L.

Low-Distortion Design [15]

In cases in which the distortion and operating temperature range constraints are relatively severe, the modified driver circuit of Figure 3.9b may become necessary. Here distortion is reduced by specifying a minimum dynamic collector current, I_0, to prevent cutoff distortion and a minimum dynamic collector voltage, V_0, to prevent saturation distortion. The dc degenerative emitter feedback provided by R_E maintains a stable quiescent point over a wide operating temperature range, which, in turn, can permit substantially the full output signal swing without saturation or cutoff distortion for the entire temperature range. The addition of R_e provides ac degenerative emitter feedback to reduce distortion and permits a larger input signal amplitude.

The dcLL for the circuit of Figure 3.9b is given by
$$V_{CC} = I_C R_C + V_{CE} + V_E, \qquad (3.22)$$

where $V_E \simeq I_C R_E$. The acLL can be expressed as

$$I_C = -(G_C + G_L)(V_{CE} - V_e) \tag{3.23}$$

in the (I_C, V_{CE}) plane. The corresponding locus of acLL midpoints is

$$I_C = (G_C + G_L)(V_{CE} - V_e), \tag{3.24}$$

which is expressed as

$$(I_C - I_0) = (G_C + G_L)(V_{CE} - V_0 - V_e) \tag{3.25}$$

when the minimum dynamic values of I_C and V_{CE} are limited to I_0 and V_0, respectively, rather than zero (Figure 3.10b). In a particular design, both $V_E \simeq I_C R_E$ and $V_e \simeq -(I_C - I_0)R_e$ should be judiciously selected to satisfy the overall constraints. Since the peak load voltage is

$$V_L = V_{CE} - V_0 - V_e, \tag{3.26}$$

(3.22) and (3.25) yield

$$I_C = \frac{V_{CC} - V_L - V_0 - V_e - V_E}{V_{CC} - 2V_L - V_0 - V_e - V_E}(V_L G_L + I_0) \tag{3.27}$$

and

$$R_C = \frac{V_{CC} - 2V_L - V_0 - V_e - V_E}{(V_L G_L + I_0)}. \tag{3.28}$$

Because the quiescent point given by (3.26) and (3.27) bisects the acLL at the nominal operating temperature T_n of the circuit, operating point drift will limit the output voltage swing to values less than V_L when $T \neq T_n$. The output voltage swing is given by

$$K_y V_L = -\frac{I_{Cy} - I_0}{G_C + G_L} \tag{3.29}$$

at the lower operating temperature limit T_y and

$$K_x V_L = V_{CEx} - V_0 - K_x V_e \tag{3.30}$$

at the upper operating temperature limit T_x where $K_y \equiv (V_L @ T_y)/(V_L @ T_n)$ and $K_x @ (V_L @ T_x)/(V_L @ T_n)$ with $0 < K_x < 1$ and $0 < K_y < 1$. As implied in Figure 3.10, the penalty for the low-distortion design is a larger quiescent current (or power) and reduced efficiency.

Although no attempt is made to do so here, the exact harmonic content or distortion of the signal can be analyzed in a conventional manner [2,16].

Transformer Coupling

Although incompatible with integrated circuit technology, transformer coupled driver stages offer improved efficiency in comparison with RC

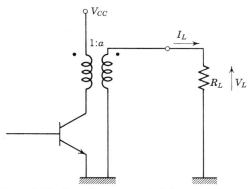

Figure 3.11 Transformer coupled class A driver stage.

coupled circuits. Assuming that transformer losses are included in the specified load power, Faraday's law of magnetic induction yields

$$\frac{V_{CC}}{1} = \frac{V_L}{a} \quad (3.31a)$$

or

$$a = \frac{V_L}{V_{CC}} \quad (3.31b)$$

for the transformer turns ratio of the circuit of Figure 3.11. From Ampere's circuital law

$$1 \cdot I_C = aI_L \quad (3.32a)$$

or

$$I_C = aI_L = a\frac{V_L}{R_L} = a^2\frac{V_{CC}}{R_L} \quad (3.32b)$$

gives the quiescent collector current. For the design described by (3.31) and (3.32), the quiescent point bisects the acLL (Figure 3.12) and the maximum

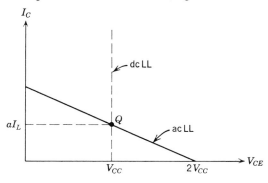

Figure 3.12 Load line diagram of class A transformer coupled driver stage.

possible efficiency of

$$\eta = \frac{P_L}{P_{DC}} = 50\% \qquad (3.33)$$

results.

3.4 OUTPUT STAGES

In order to minimize the quiescent power of a low-frequency amplifier, it is particularly advantageous to use an output circuit configuration especially adapted for minimum standby power drain. The class B push-pull output

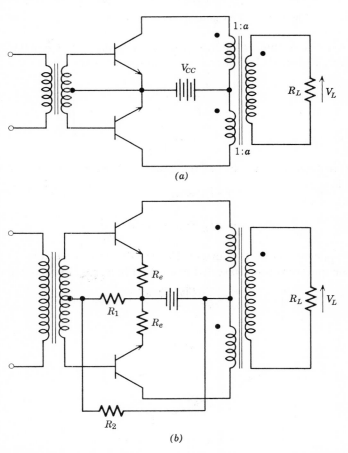

Figure 3.13 Transformer coupled push-pull output stage: (*a*) class B, (*b*) class AB with ac emitter feedback.

stage illustrated in Figure 3.13 is quite suitable for this purpose. The primary design constraints for this circuit are similar to those of the driver stage.

Class B Transformer Coupled Circuits

The push-pull common emitter output stage [1,2,3,16,17] in Figure 3.13a dissipates an extremely small quiescent power

$$P_{DC} \simeq 2V_{CC}I_{CBS} \qquad (3.34)$$

that depends on I_{CBS}, the reverse saturation current of the collector junction with $V_{BE} \simeq 0$. The load line diagram for the circuit is illustrated in Figure 3.14. The design proceeds as described for the class A transformer coupled driver stage so that $a = V_L/V_{CC}$ and $I_C = aI_L$ is the maximum collector current. The maximum collector voltage is $V_{CE} = 2V_{CC}$. The maximum possible efficiency is

$$\eta = \frac{P_L}{P_{DC}} \simeq \frac{\tfrac{1}{2}V_L I_L}{V_{CC}(2I_C/\pi)} = 78\%. \qquad (3.35)$$

Since

$$P_{DC} = V_{CC}\frac{I_C}{\pi}\int_0^\pi \sin\phi\, d\phi = V_{CC}\left(\frac{2I_C}{\pi}\right)$$

where $\phi = \omega t$.

The location of the quiescent point Q at $(0, V_{CC})$ creates a distortion problem in the crossover range of operation of the circuit of Figure 3.13a. Crossover distortion is caused principally by the nonlinear input resistance of the transistor and the operating point dependence of its current gain. This problem can be alleviated by providing a small forward bias voltage for the emitter junction as well as an emitter feedback resistor as illustrated in Figure 3.13b. However, the resulting class AB operation entails a larger quiescent power, and both R_1 and R_e tend to reduce the power gain of the circuit.

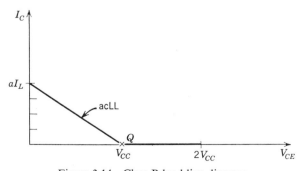

Figure 3.14 Class B load line diagram.

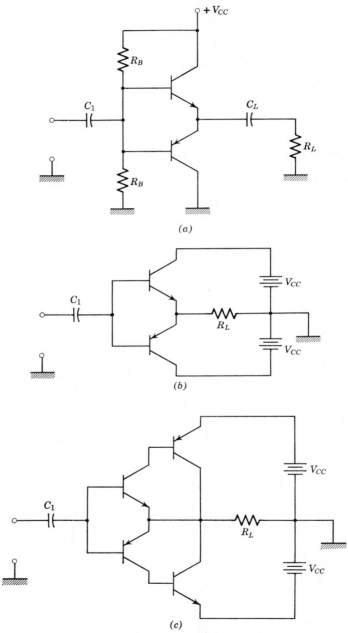

Figure 3.15 Complementary symmetry class B output stages.

Class B Complementary Symmetry Circuits

The input and output transformers of the circuits of Figure 3.13 can be eliminated by using complementary symmetry circuits [1,2,18] as illustrated in Figure 3.15. In these circuits the peak load voltage is $V_L = V_{CC}$ and the peak load current is $I_L = I_C$, where I_C may be taken as the sum of the two collector currents for the circuit of Figure 3.15c. The emitter follower connection of these complementary circuits provides ac degenerative feedback, which reduces distortion at the sacrifice of power gain. Providing (a) a slight forward bias for the emitter junctions, (b) an overall feedback loop including the driver and possibly earlier stages [19], or (c) a high-impedance driver that substantially reduces the crossover distortion of the transistor [20] tends to alleviate the distortion problem.

A Monolithic Hearing Aid Amplifier

A micropower low-frequency amplifier that has been fabricated as a monolithic integrated circuit [21] is illustrated in Figure 3.16. The circuit is intended

Table 3.1 Characteristics of Monolithic Hearing Aid Amplifier

Operating voltage	1.1–1.55	volts
Nominal load impedance	1000	ohms
Source impedance	5000	ohms
Temperature range	0–50	°C
Volume control	0–4	kilohms (nonlinear)
Performance (1.55 V)		
Power output	3 mW	min
Harmonic distortion	5 %	max
Current at 3 mW	4.5 mA	max
Idling current	1.0 mA	typ
Voltage gain	72 dB	min
Input impedance	40 Kohms	typ
Volume control range	35 dB	min
Noise	3 mV	typ
Stable with 50 ohms battery resistance		

for use as a hearing aid amplifier. Its salient characteristics are listed in Table 3.1. The circuit is essentially a balanced class A preamplifier driving a class B push-pull output stage. Direct coupling is used throughout, and

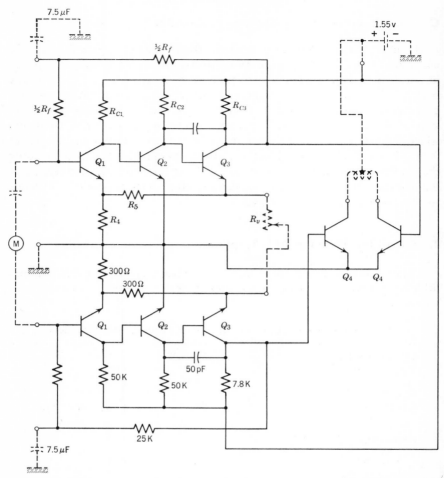

Figure 3.16 Class B silicon monolithic integrated circuit hearing aid amplifier (after Hellstrom [21]).

separate ac and dc feedback paths are employed. Gain control is accomplished by means of an external variable resistance R_v. When R_v is zero, the out of phase currents in R_4 and R_5 from the six class A preamplifier transistors cancel and the emitters are effectively at ac ground. When R_v is maximum, the two channels operate independently and negative ac feedback is introduced by R_4 and R_5. The dc feedback loop acts to maintain the base of Q_4 at a potential sufficiently above the base of Q_1 to supply the base current required through R_f.

A Monolithic Acoustic Horn Amplifier

Another micropower low-frequency amplifier that has been fabricated as a monolithic integrated circuit [22] is illustrated in Figure 3.17. The circuit is intended for use as an audio amplifier for an acoustic horn speaker used in an FM helmet radio receiver [22]. Its salient properties are

power gain	40 dB
input impedance	20 K
power drain	20 mW at $P_L = 5$ mW
power drain	4.5 mW at $P_L = 0$
supply voltage	2.2–3.0 V
distortion	4% at $P_L = 5$ mW, $f = 1$ kHz

The circuit consists of a balanced class A preamplifier that is a direct coupled common-collector, common-emitter cascade. The emitter follower driver stage receives its collector voltage from the squelch switch, the output stage of which is essentially an inverter that is cutoff in the receive mode of operation and saturated in the standby mode. The class B common-emitter output stage drives a center-tapped acoustic horn speaker.

3.5 CLASS D OR TWO-STATE AMPLIFIERS

As a consequence of its low-standby power drain and high efficiency, the class B amplifier is attractive for micropower applications as a push-pull output stage. The normal class C amplifier, in which a transistor remains cutoff for more than half the period of the input signal, presents severe crossover distortion problems that preclude its use as a substitute for the class B stage. However, a class D or two-state amplifier [23–28] can provide improved efficiency and adequate fidelity compared with the push-pull class B circuit and, consequently, is also attractive for use in micropower applications.

A class D amplifier is one in which the transistor is used as a switch, being either turned on (full current at nearly saturation voltage) or off (nearly zero current at full voltage), dissipating significant power only during the switching transitions, which can be a small fraction of the signal period. Amplification of a signal is achieved by rapidly switching an inductive load between a positive and a negative voltage. Such a pulse train is shown in Figure 3.18. With the duty cycle at 50% (that is, the time the load is connected to one side of the supply voltage is 50% of one period), the average value of this waveform is zero and is shown by the dotted line in Figure 3.18a. If the duty cycle

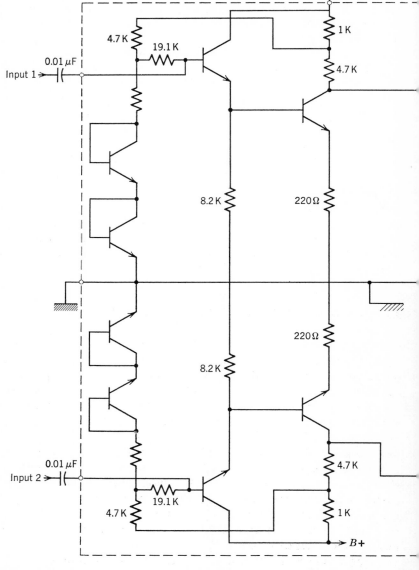

Figure 3.17 Class B silicon monolithic integrated circuit acoustic horn amplifier (after Gilson and Quaid [22]).

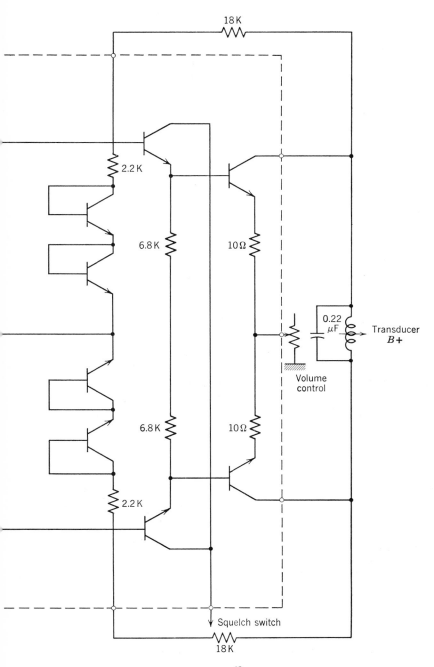

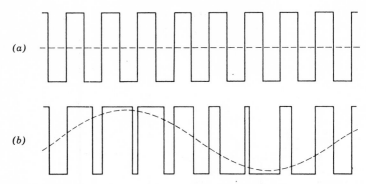

Figure 3.18 Output waveforms produced by a class D amplifier: (*a*) unmodulated; (*b*) width-modulated.

is altered, as shown in Figure 3.18*b*, the average voltage will change proportionally. The inductor in series with the load, acting as a low-pass filter, rejects the high frequencies of the pulse train, but passes the average voltage. If the switching rate is rather high (for instance, 100 kHz) and the waveform is fed through an inductor into a resistive load, the average voltage (for example, a sine wave) will develop across the load.

In a class D amplifier, the signal is converted into a width-modulated pulse train which is amplified and finally demodulated. Figure 3.19 illustrates a simple pulse width modulator using a high-frequency sawtooth waveform. Figure 3.20 illustrates two-output stage configurations. An efficiency in excess of 85%, a frequency response of 0 to 15 kHz and a distortion of less than 1% at 1 watt output have been achieved in complementary class D amplifiers [23].

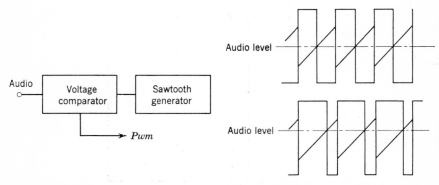

Figure 3.19 Simple pulse width modulator (after Camenzend [23]).

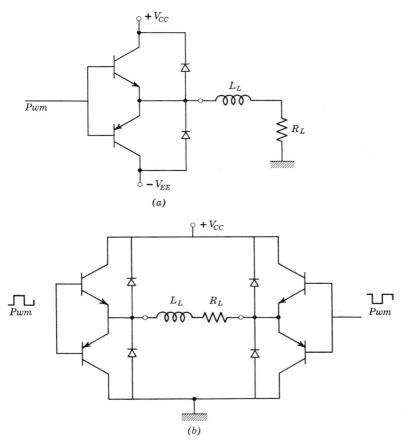

Figure 3.20 Class D output stages. (*a*) complementary emitter follower; (*b*) complementary bridge.

REFERENCES

[1] A. W. Lo et al., *Transistor Electronics*, Prentice-Hall, Englewood Cliffs, N.J., 1955, Chapter 5.
[2] R. F. Shea et al., *Transistor Circuit Engineering*, Wiley, New York, 1957, Chapter 4.
[3] D. Dewitt and A. L. Rossoff, *Transistor Electronics*, McGraw-Hill, New York, 1957, Chapter 8.
[4] J. G. Linvill and J. F. Gibbons, *Transistors and Active Circuits*, McGraw-Hill, New York, 1961, Chapter 10.
[5] E. Keonjian, "Micropower Audio Amplifier," *IRE Trans. Circuit Theory*, **3**, 56 (March 1957).
[6] E. Keonjian, "Micropower Operation of Silicon Transistors," *Tele-Tech Electron. Ind.*, **15**, 76–79 (May 1956).

[7] C. D. Todd, "Micropower Preamplifier," *Electron. Equip. Eng.*, **8**, 41–44 (June 1960).
[8] W. W. Gaertner et al., "Microelectronic, Micropower Digital Circuits and Low Level Amplifiers for Space Applications," Spaceborne Computer Engineering Conference, Anaheim, Calif. (October 1962).
[9] W. W. Gaertner et al., "Micropower Microelectronic Subsystems," in *Micropower Electronics*, E. Keonjian, Ed., Macmillan, New York, 1964, pp. 57–84.
[10] R. D. Middlebrook and C. A. Mead, "Transistor AC and DC Amplifiers with High Input Impedance," *Semicond. Prod.*, **2**, 26–35 (March 1959).
[11] J. A. Ekiss, "Transistor ac Amplifiers with High Input Impedance: A Survey," *J. Audio Engineering Soc.*, **8**, 18–22 (January 1960).
[12] L. J. Sevin, *Field-Effect Transistors*, McGraw-Hill, New York, 1965, Chapter 3.
[13] J. S. Sherwin, "An FET Micropower Amplifier," *Electronics*, **37**, 74–75 (December 14, 1964).
[14] J. D. Meindl and P. H. Hudson, "Low Power Linear Circuits," *IEEE J. Solid-State Circuits*, **1**, 100–111 (December 1966).
[15] J. D. Meindl et al., "Static and Dynamic Performance of Micropower Transistor Linear Amplifiers," in *Micropower Electronics*, E. Keonjian, Ed., Macmillan, New York, 1964.
[16] W. W. Gaertner, *Transistors: Principles, Design and Applications*, Van Nostrand, Princeton, N.J., 1960, Chapter 13.
[17] H. C. Lin, "Tradeoffs in Selecting Low-Power Integrated Circuits," *Electron. Products*, **7**, 32–36 (August 1965) and **7**, 28–32 (September 1965).
[18] L. P. Hunter, *Handbook of Semiconductor Electronics*, 2nd Ed., McGraw-Hill, New York, 1962, Chapter 11.
[19] J. B. Izatt, "Class-AB Amplifiers Employing a Complementary Pair of Transistor," *Proc. IEE, British* **113**, 721–724 (May 1966).
[20] J. J. Faran and R. G. Fulks, "High-Impedance Drive for the Elimination of Crossover Distortion," *Solid-State J.*, **2**, 36–40 (August 1961).
[21] M. J. Hellstrom and J. J. Hsieh, "A Monolithic Silicon Class B Hearing Aid Amplifier," Session 1, 1965 WESCON, San Francisco, Calif., August 24–27, 1965.
[22] R. A. Gilson and T. B. Quaid, "An Integrated Circuit Helmet Radio Receiver," IEEE National Conv., New York, March 1967.
[23] H. R. Camenzend, "Modulated Pulse Audio and Servo Power Amplifiers," *Digest of Technical Papers*, 1966 International Solid-State Circuits Conference, University of Pennsylvania, Philadelphia, Pa. (February 1966), pp. 90–91.
[24] D. F. Page et al., "On Solid-State Class-D Systems," *Proc. IEEE*, **53**, 423–424 (April 1965).
[25] M. L. Stephens and J. P. Wittman, "Switched Mode Transistor Amplifiers," *IEEE Trans. Commun. Electron.*, **82**, 470–472 (September 1963).
[26] Amar G. Base, "A Two-State Modulation System," *Electro-Technol.*, **74**, 42–47 (August 1964).
[27] A. W. Carlson et al., "Linear Power Amplification Using Switching Techniques," *Dig. Tech. Papers*, 1965 NEREM, Boston, Mass., **VII**, 86–87 (November 1965).
[28] E. C. Bell and T. Sergent, "Distortion and Power Output of Pulse Duration Modulated Amplifiers," *Electron. Eng.*, **37**, 540–542 (August 1965).

Chapter 4

Wideband Amplifiers

The design constraints for a wideband amplifier [1–10] may include gain, bandwidth, phase shift, input and output impedance, noise figure, roll-off behavior, dynamic range, number of stages, operating temperature range, and supply voltage as well as other requirements. In most instances the principal factors that determine the quiescent power of the amplifier are gain, bandwidth, input impedance, and noise figure for an input stage; gain and bandwidth for intermediate stages; and gain, bandwidth, output impedance, and dynamic range for an output stage. The problem of interest in this chapter is the design of micropower wideband amplifiers for minimum quiescent power drain.

4.1 INPUT STAGES

The input impedance of the wideband common emitter input stage illustrated in Figure 4.1 is

$$Z_{ie} = \frac{V_{in}}{I_{in}} \simeq \frac{r_{be} \parallel R_p}{1 + j\omega(r_{be} \parallel R_p)[C_{be} + (1 + A_{v0})C_c]}, \qquad (4.1)$$

where $r_{be} \parallel R_p = r_{be}R_p/(r_{be} + R_p)$ and $A_{v0} = g_m R_L$ is the midband voltage amplification of the stage whose ac load resistance is R_L. The midband

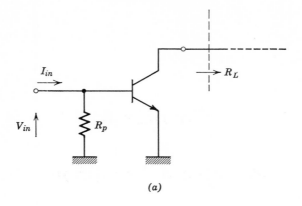

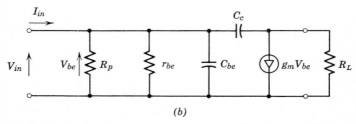

Figure 4.1 Wideband amplifier input stage: (*a*) ac schematic diagram; (*b*) equivalent circuit.

magnitude R_{ie} and bandwidth ω_{ie} of Z_{ie} are

$$R_{ie} = \frac{r_{be} R_p}{r_{be} + R_p} \qquad (4.2)$$

and

$$\omega_{ie} = \frac{1}{R_{ie}[C_{be} + (1 + A_{v0})C_c]}. \qquad (4.3)$$

On the basis of an overall analysis of the amplifier, frequently A_{v0} and R_L are designated for the input stage, and $g_m = A_{v0}/R_L$ determines I_C. Consequently, the value of R_p necessary to produce the specified R_{ie} can be calculated from (4.2) and the corresponding value of ω_{ie} obtained from (4.3). Using a high-performance micropower transistor (for which ω_T (max) $\gg \omega_{ie}$), typically, the input stage can be designed so that $r_{be} \gg R_p \simeq R_{ie}$ to provide the specified value of R_{ie} and an ω_{ie} that is in excess of its specified value. In those instances where the required value of R_{ie} exceeds $r_{be} = h_{fe0}/g_m = h_{fe0}(R_L/A_{v0})$, from (4.2), or as the product of the specified values of

R_{ie} and ω_{ie} exceeds $[C_{be} + (1 + A_{v0})C_c]^{-1}$, from (4.3), an ac emitter feedback resistor or an emitter follower configuration can be utilized to increase R_{ie}.

4.2 INTERMEDIATE STAGES

The primary design constraints that limit the minimum quiescent power of the intermediate stages of a micropower wideband amplifier are overall midband amplification A_{v0}^* and upper 3 dB cutoff frequency or bandwidth, ω_c^* [11].

Bipolar Transistor Circuits

Figure 4.2 illustrates the circuit diagram for a micropower wideband iterative amplifier. A key simplification in the analysis of this amplifier is the development of the unilateral iterative stage equivalent circuit illustrated in Figure 4.2c [5,6,11,12]. Consider the equivalent circuit of Figure 4.3, in which the load of the micropower iterative stage is approximated as a fixed parallel resistance R_L and capacitance C_L. This assumption is certainly valid over a limited frequency band. The impedance that is seen looking into the collector capacitance is

$$Z_m = \frac{V_{be}}{I_m} = \frac{1}{1 + g_m Z_L}\left(Z_L + \frac{1}{j\omega C_c}\right), \quad (4.4)$$

where $Z_L = R_L/(1 + j\omega R_L C_L)$. In most instances for a micropower wideband amplifier, $|Z_L| \ll |1/j\omega C_c|$; that is, the load on the output current generator $g_m V_{be}$ is principally Z_L as opposed to C_c. Consequently

$$Z_m \simeq \frac{1}{j\omega(1 + g_m Z_L)C_c}, \qquad |Z_L| \ll |1/j\omega C_c|. \quad (4.5)$$

Since $V_L = (I_m - g_m V_{be})Z_L \simeq -g_m V_{be} Z_L$, if the forward transmission I_m through C_c is much less than the output current generator $g_m V_{be}$, then (4.5) may be written as

$$Z_m \simeq \frac{1}{j\omega(1 + A_v)C_c} \quad (4.6)$$

$$|I_m| \ll g_m V_{be} \quad \text{and} \quad A_v = -\frac{V_L}{V_{be}} = g_m Z_L.$$

Now, *solely* for the purpose of calculating Z_m, it is assumed that

$$A_v = g_m Z_L \simeq g_m R_L = A_{v0}$$

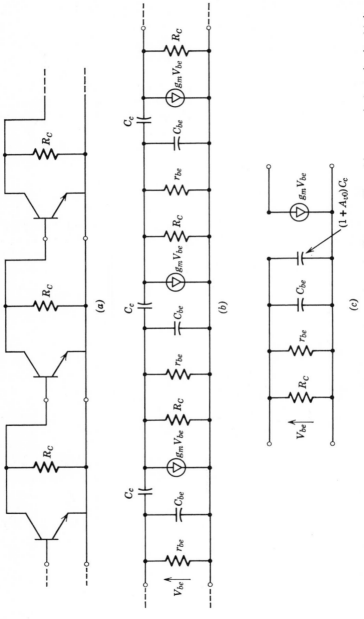

Figure 4.2 Wideband iterative amplifier circuit diagrams: (a) ac schematic diagram; (b) multistage equivalent circuit; (c) iterative stage equivalent circuit.

Intermediate Stages

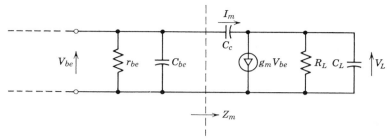

Figure 4.3 Model used to form unilateral equivalent circuit of Figure 4.2c.

or that the load impedance, $Z_L \simeq R_L$, is entirely real. Therefore

$$Z_m \simeq \frac{1}{j\omega(1 + A_{v0})C_c} \tag{4.7}$$

permits the unilateral iterative stage equivalent circuit of Figure 4.2c.

The voltage gain (or current gain) of the iterative stage is

$$-\frac{v_n}{v_{n-1}} = A_v = \frac{A_{v0}}{1 + j\omega(A_{v0}/g_m)[C_{be} + (1 + A_{v0})C_c]}, \tag{4.8}$$

where the midband gain is

$$A_{v0} = h_{fe0}\frac{R_c}{R_c + h_{fe0}/g_m} \tag{4.9}$$

and the bandwidth is

$$\omega_c = \frac{g_m}{A_{v0}[C_{be} + (1 + A_{v0})C_c]}. \tag{4.10}$$

From (4.10), the gain-bandwidth product

$$A_{v0}\omega_c = \frac{g_m}{C_{be} + (1 + A_{v0})C_c} \tag{4.11}$$

of a micropower wideband iterative stage is directly proportional to the quiescent collector current, since $g_m = qI_C/kT$, and inversely proportional to gain A_{v0} through the Miller effect capacitance $(1 + A_{v0})C_c$.

The overall voltage gain of the n stage amplifier is

$$A_v^* = A_v^{\,n} = \left[\frac{A_{v0}}{1 + j\omega/\omega_c}\right]^n, \tag{4.12}$$

which gives

$$A_{v0}^* = A_{v0}^{\,n}, \qquad \omega \ll \omega_c \tag{4.13}$$

in the midband range and

$$\left|\frac{A_v^*}{A_{v0}{}^n}\right| = \frac{1}{\sqrt{2}} = \left|\frac{1}{(1 + j\omega_c^*/\omega_c)^n}\right| \tag{4.14}$$

or

$$\omega_c^* = \sqrt{2^{1/n} - 1}\,\omega_c \simeq \frac{\omega_c}{1.2\sqrt{n}} \tag{4.15}$$

as the overall 3 dB cutoff frequency.

The overall quiescent power drain for a typical micropower wideband circuit (see Figure 2.4 or 2.5) is

$$P_{DC}^* = nP_{DC} \simeq nV_{CC}I_C \tag{4.16a}$$

or

$$P_{DC}^* \simeq n[I_C R_C + V_{CE}]I_C. \tag{4.16b}$$

Combining (4.9) and (4.10) with (4.16b) gives

$$P_{DC}^* = \gamma V_{CE}\omega_c C_c \left[\frac{\gamma}{V_{CE}} \frac{A_{v0}}{1 - A_{v0}/h_{fe0}} + 1\right] nA_{v0}\left[\frac{C_{be}}{C_c} + 1 + A_{v0}\right], \tag{4.17}$$

which when combined with (4.13) and (4.15) and normalized yields [11]

$$\frac{P_{DC}^*}{1.2\gamma V_{CE}\omega_c^* C_c} = \left[\frac{\gamma}{V_{CE}} \frac{(A_{v0}^*)^{1/n}}{1 - (A_{v0}^*)^{1/n}/h_{fe0}} + 1\right] n^{3/2}(A_{v0}^*)^{1/n}\left[\frac{C_{be}}{C_c} + 1 + (A_{v0}^*)^{1/n}\right]. \tag{4.18}$$

This result indicates the following for minimum power drain, P_{DC}^*, in a micropower wideband amplifier:

1. The transistor capacitances C_{be} and C_c, which limit f_T, should be as small as possible.
2. The transistor small signal current gain should be much larger than the midband gain per stage (that is, $h_{fe0} \gg A_{v0} = (A_{v0}^*)^{1/n}$).
3. The collector-to-emitter voltage, V_{CE}, of the transistor should be as small as possible when V_{CC} is adjustable as (4.16b) presumes.
4. The number of stages, n, in the amplifier should be adjusted to the optimum value.

Notice also that total power drain is directly proportional to the overall bandwidth ω_c^* of the amplifier.

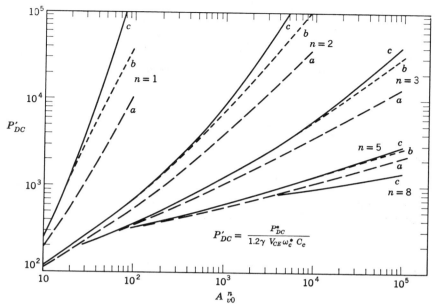

Figure 4.4 Normalized quiescent power P'_{DC} for a wideband iterative amplifier versus overall amplification $A_{v0}^* = A_{v0}{}^n$ for various numbers of stages n.

Figure 4.4 displays a plot of (4.18) for $C_{be}/C_c = 8$, $h_{fe0} = 120$, and $V_{CE} = 3.0$ V. The a curves reflect only transistor quiescent power, which is described by (4.18) for

$$\frac{\gamma}{V_{CE}} \frac{A_{v0}}{1 - A_{v0}/h_{fe0}} \ll 1.$$

The b curves reflect circuit quiescent power P_{DC}^* for $A_{v0} \ll h_{fe0}$. The c curves reflect the increase in circuit power for typical designs where $A_{v0} \to h_{fe0}$. As Figure 4.4 indicates, by increasing the number of stages of a wideband iterative amplifier beyond what might be considered the minimum number for a fixed A_{v0}^*, a very substantial quiescent power savings is possible. Such increases in the number of stages are most attractive when considered in terms of the batch fabrication technologies of integrated electronics.

If V_{CC} is fixed, combining (4.10), (4.11), and (4.15) with (4.16a) gives

$$\frac{P_{DC}^*}{1 \cdot 2\gamma V_{CC} \omega_c^* C_c} = n^{3/2}(A_{v0}^*)^{1/n}\left[\frac{C_{be}}{C_c} + 1 + (A_{v0}^*)^{1/n}\right]. \quad (4.19)$$

For a normalized power

$$P'_{DC} = \frac{P_{DC}^*}{1 \cdot 2\gamma V_{CC} \omega_c^* C_c}, \quad (4.20)$$

the a curves of Figure 4.4 describe the circuit quiescent power requirements of (4.19). For example, the normalized quiescent power required for an overall gain $A_{v0}^* = 2500$ can be reduced from $P_{DC}' = 8350$ for $n = 2$ (and $A_{v0} = 50$) to $P_{DC}' = 742$ for $n = 5$ (and $A_{v0} = 4.78$). From (4.19) the optimum value for n can be found by taking the derivative $\partial P_{DC}'/\partial n$, which yields

$$n = \frac{2}{3}\left[1 + \frac{A_{v0}}{\frac{C_{be}}{C_c} + 1 + A_{v0}}\right] \ln A_{v0}^*. \tag{4.21}$$

For $C_{be}/C_c \gg A_{v0} > 1$, (4.21) gives

$$n = \tfrac{2}{3} \ln A_{v0}^* \quad \text{and} \quad A_{v0} = \epsilon^{3/2} = 4.50 \tag{4.22}$$

as the optimum number of stages and gain per stage, respectively, for minimum P_{DC}'. For $1 < C_{be}/C_c \ll A_{v0}$ the corresponding results are

$$n = \tfrac{4}{3} \ln A_{v0}^* \quad \text{and} \quad A_{v0} = \epsilon^{3/4} = 2.12. \tag{4.23}$$

Consequently the optimum values of n and A_{v0} range between

$$\tfrac{2}{3} \ln A_{v0}^* \leq n \leq \tfrac{4}{3} \ln A_{v0}^* \tag{4.24}$$

and $6.54 \text{ dB} = 2.12 = \epsilon^{3/4} \leq A_{v0} \leq \epsilon^{3/2} = 4.50 = 13.1 \text{ dB}$ for minimum quiescent power.

Figure 4.5 provides a further illustration of the influence of the number of stages of a micropower wideband amplifier on the quiescent power requirements of the amplifier. Based on (4.18) and, in essence, constituting merely a modified presentation of the data in Figure 4.4, the curves in Figure 4.5 clearly indicate (a) the existence of an optimum value for n at which P_{DC}^* is minimum, and (b) the fact that, in most instances, to achieve a judicious compromise between the amount of hardware or number of stages and the quiescent power, it is advisable to select a value of n somewhat smaller than the optimum.

A graphical summary of the more prominent features of a micropower wideband iterative stage is provided by Figure 4.6. The design begins with the selection of a transistor whose maximum gain-bandwidth product is well beyond the range of direct interest for the circuit under consideration (for example, $f_T = 1000$ MHz at $I_C = 1.0$ mA). The quiescent collector current of the transistor is then reduced until its gain-bandwidth product exactly meets the performance constraints of the circuit (for example, $f_T = 10$ MHz at $I_C = 10 \mu\text{A}$). This provides a substantial reduction in quiescent current (or power); however, the cutoff frequency of h_{fe} (that is, $f_\beta = f_T/h_{feo}$) is much smaller (for instance, $f_\beta \simeq 100$ kHz) than the required bandwidth of the stage (for instance, $\omega_c/2\pi = f_c \simeq 1.0$ MHz). Consequently the value of the

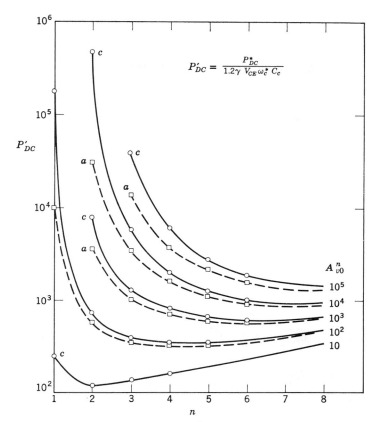

Figure 4.5 Normalized quiescent power P'_{DC} for a wideband iterative amplifier versus the number of stages n for various overall amplifications $A^*_{v0} = A_{v0}^n$.

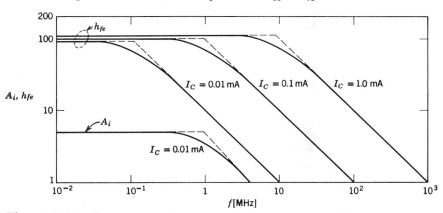

Figure 4.6 Transistor current gain, h_{fe}, and wideband iterative stage current gain, $A_i = A_v$, versus frequency.

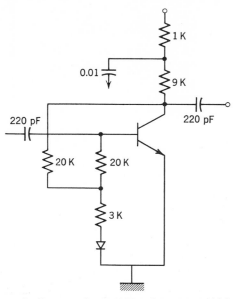

Figure 4.7 Schematic diagram of a 455 kHz micropower wideband amplifier stage.

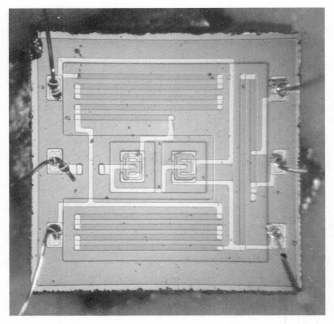

Figure 4.8 Microphotograph of a micropower wideband amplifier stage fabricated as a monolithic integrated circuit. Resistor line width is 1 mil (courtesy of U.S. Army Electronics Command).

interstage resistor, $R_c \ll r_{be}$, is chosen to extend the iterative stage bandwidth to the required value while drastically reducing the stage gain (for example, $A_{v0} \simeq 5 \ll h_{fe0} \simeq 100$). Since the amplifier bandwidth, ω_c^*, is proportional to $n^{-1/2}$ while amplification, A_{v0}^*, is proportional to $A_{v0}{}^n$, we can profitably add several micropower stages to provide the specified A_{v0}^* and ω_c^*. Techniques for broadbanding [7,8,9] such as shunt peaking and series peaking can provide additional small reductions in the quiescent power of a wideband micropower amplifier.

Figure 4.7 illustrates the schematic diagram of a micropower wideband amplifier stage that was designated for use in a portable FM receiver [13] intended for military applications. A photograph of one of these stages fabricated as a silicon monolithic integrated circuit is illustrated in Figure 4.8. All capacitors are conventional discrete components. The salient features of the performance of a single stage are listed:

voltage gain	20 dB
cutoff frequency	455 kHz
supply voltage	3.0 V
quiescent power	450 μW

FET Circuits [14]

The circuit diagrams of a micropower FET wideband iterative amplifier are illustrated in Figure 4.9. Following the pattern of (4.4)–(4.8), the basis for the iterative stage equivalent circuit of Figure 4.9c can be established.

The voltage gain of the iterative stage is

$$A_v = \frac{A_{v0}}{1 + j\omega(A_{v0}/g_{m0})[C_{gs} + C_{ds} + (1 + A_{v0})C_{gd}]}, \quad (4.25)$$

where the midband gain is

$$A_{v0} \simeq g_{m0} R_D \quad (4.26)$$

and the bandwidth is

$$\omega_c = \frac{g_{m0}}{A_{v0}[C_{gs} + C_{ds} + (1 + A_{v0})C_{gd}]}. \quad (4.27)$$

Since

$$g_m = \frac{\partial I_D}{\partial V_{GS}} = \frac{2I_{DSS}}{V_P}\left(1 - \frac{V_{GS}}{V_P}\right), \quad (4.28)$$

to achieve minimum quiescent power (or I_{DSS}) it is assumed that the FET operates at $V_{GS} = 0$ so that in (4.25)–(4.27)

$$g_{m0} = g_m\bigg|_{V_G=0} = \frac{2I_{DSS}}{V_P}. \quad (4.29)$$

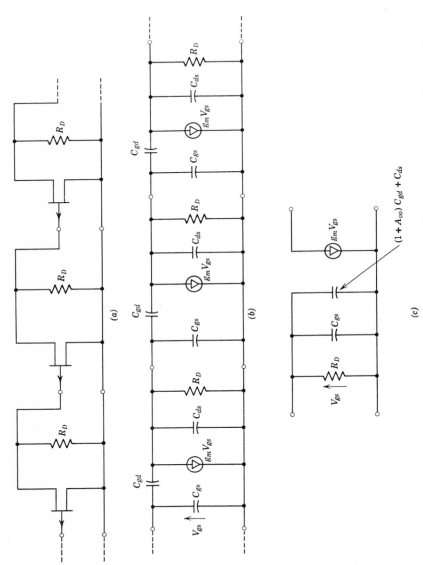

Figure 4.9 Wideband FET iterative amplifier circuit diagrams: (a) ac schematic diagram; (b) multistage equivalent circuit; (c) iterative stage equivalent circuit.

Intermediate Stages

An important implication of (4.28) is that the geometry of an FET, most conveniently the channel width W, has an optimum value for which the quiescent power of a specified wideband amplifier is a minimum. Since

$$A_v^* = A_v^n = \left(\frac{A_{v0}}{1 + j\omega/\omega_c}\right)^n \tag{4.30}$$

holds for the FET amplifier, following (4.11)–(4.14) for the bipolar transistor circuit, the overall quiescent power of the FET amplifier is

$$P_{DC}^* \simeq nV_{CC}I_{DSS} \tag{4.31a}$$

$$\simeq n(I_{DSS}R_D + V_{DS})I_{DSS}, \tag{4.31b}$$

which combined with (4.26) and (4.29) and normalized yields [11]

$$\frac{P_{DC}^*}{0.6V_P V_{DS}\omega_c^* C_{gd}} = \left[\frac{V_P}{2}\frac{(A_{v0}^*)^{1/n}}{V_{DS}} + 1\right]n^{3/2}(A_{v0}^*)^{1/n}\left[\frac{C_{gs} + C_{ds}}{C_{gd}} + 1 + (A_{v0}^*)^{1/n}\right]. \tag{4.32}$$

For minimum P_{DC}^* in a micropower FET wideband amplifier, this result indicates the following:

1. The FET capacitances C_{gs}, C_{gd}, and C_{gs} should be as small as possible.
2. The pinch-off voltage V_P should be as small as possible.
3. The drain-to-source voltage V_{DS} should be as small as possible when V_{CC} is adjustable as (4.31b) presumes.
4. The number of stages n should be adjusted to the optimum value.
5. The FET should be operated at zero gate-to-source voltage $V_{GS} = 0$, which is implicit in (4.32).

Figure 4.10 displays a plot of (4.31) for $(C_{gs} + C_{ds})/C_{gd} = 3$, $V_P = 2.0$ V and $V_{DS} = 5.0$ V. The dashed curves represent only FET quiescent power, which is described by (4.31) for

$$\frac{V_P}{2}\frac{1}{V_{DS}}A_{v0} \ll 1.$$

The solid curves reflect the circuit quiescent power. As Figure 4.10 indicates, by increasing the number of stages of a wideband FET iterative amplifier beyond what might be considered the minimum number for a fixed A_{v0}^*, a most substantial quiescent power savings is possible.

If V_{CC} is fixed, combining (4.27) and (4.30) with (4.30a) gives

$$\frac{P_{DC}^*}{0.6V_P V_{CC}\omega_c^* C_{gd}} = n^{3/2}(A_{v0}^*)^{1/n}\left[\frac{C_{gs} + C_{ds}}{C_{gd}} + 1 + (A_{v0}^*)^{1/n}\right]. \tag{4.33}$$

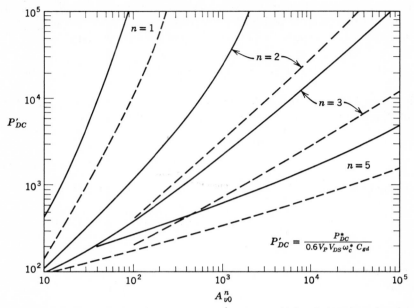

Figure 4.10 Normalized quiescent power P'_{DC} for a wideband FET iterative amplifier versus overall amplification $A^*_{v0} = A_{v0}{}^n$ for various numbers of stages n.

For a normalized power

$$P'_{DC} = \frac{P^*_{DC}}{0.6 V_P V_{CC} \omega^*_c C_{gd}} \quad (4.34)$$

the dashed curves in Figure 4.10 describe the circuit quiescent power requirements of (4.33); for example, the normalized quiescent power required for an overall gain $A^*_{v0} = 2500$ can be reduced from $P'_{DC} = 7650$ for $n = 2$ (and $A_{v0} = 50$) to $P'_{DC} = 473$ for $n = 5$ (and $A_{v0} = 4.78$). As a consequence of the similar form of (4.19) and (4.33), it can be shown that conditions equivalent to (4.21)–(4.23) are characteristic of the FET amplifier. Since $C_{be}/C_c > (C_{gs} + C_{ds})/C_{gd}$ in most cases, we would expect a somewhat larger optimum number of stages (and therefore a smaller gain per stage) for the FET amplifier. A comparison of (4.19) and (4.33) gives

$$\frac{P^*_{DC}(\text{FET})}{P^*_{DC}(\text{bipolar})} = \frac{1}{2} \frac{V_P}{\gamma} \frac{C_{gd}}{C_c} \frac{[(C_{gs} + C_{ds})/C_{gd} + 1 + A_{v0}]}{[C_{be}/C_c + 1 + A_{v0}]} \quad (4.35)$$

for equal V_{CC}, A^*_{v0}, ω^*_c, and n. For a rough approximation,

$$\frac{P^*_{DC}(\text{FET})}{P^*_{DC}(\text{bipolar})} \simeq \frac{1}{2} \frac{V_P}{\gamma} = \frac{1}{2} \frac{V_P}{kT/q} \quad (4.36)$$

reveals that for $1.0 \text{ V} \leq V_P \leq 2.0 \text{ V}$, the quiescent power required by an FET amplifier is 20 to 40 times as large as the power of a comparable bipolar transistor circuit. Since

$$\frac{P_{DC}^*(\text{FET})}{P_{DC}^*(\text{bipolar})} = \frac{g_m(\text{bipolar})}{I_C} \cdot \frac{I_{DSS}}{g_{m0}(\text{FET})}, \quad (4.37)$$

it is evident that the different transconductances of the two types of transistors lead directly to the different quiescent power requirements of the circuits. Disregarding quiescent current, the bipolar transistor transconductance

$$g_m = \frac{q}{kT} I_C \quad (4.38)$$

depends only on physical constants, whereas the FET transconductance (4.29) depends on device geometry and material properties, which, in turn, are constrained by the limitations of the process technology used in device fabrication.

4.3 OUTPUT STAGES

The design constraints for a micropower wideband output stage may include supply voltage V_{CC}, dynamic range or peak load voltage V_L, load impedance Z_L, output impedance Z_0, voltage gain A_{v0}, and bandwidth ω_c. Several typical sets of constraints and the corresponding design procedures are discussed briefly in this section.

With V_{CC}, V_L, and $Z_L^{-1} = 1/R_L + j\omega C_L$ specified for the output stage in Figure 4.11, the minimum quiescent power design is described by (3.19)

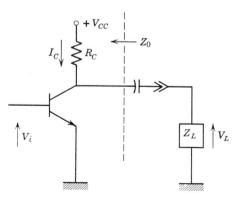

Figure 4.11 Wideband amplifier output stage.

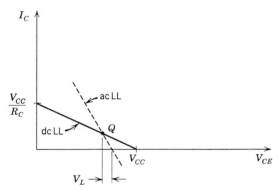

Figure 4.12 Output stage load line diagram for V_{CC}, V_L, R_L, and R_C specified.

and (3.20). With I_C and R_C established, the midband voltage gain and bandwidth of the output stage can be calculated directly from

$$-\frac{V_L}{V_i} = A_v \simeq \frac{g_m}{1/R_C + 1/R_L + j\omega C_L} \qquad (4.39)$$

to determine their compatibility with the design objectives.

With V_{CC}, V_L, Z_L, and $Z_0 \simeq R_C$ specified, both the dcLL and acLL of Figure 4.12 are fixed, and minimum power requires that I_C take the value just large enough to permit the required $V_L < V_{CE}$. From Figure 4.12, the quiescent point Q is located at

$$I_C = (G_C + G_L)V_L \qquad (4.40a)$$

and

$$V_{CE} = V_{CC} - \frac{I_C}{G_C} = V_{CC} - \left(\frac{G_C + G_L}{G_C}\right)V_L. \qquad (4.40b)$$

Again, A_{v0} and ω_c can be determined from (4.39).

With V_{CC} and R_L specified as well as an additional requirement for symmetrical limiting an optimum design for minimum power exists [1]. Solving the dc load equation (3.16) and the equation for the locus of acLL midpoints (3.18) for I_C and using $V_{CE} = V_L$ give

$$\eta = \frac{P_L}{P_{DC}} = \frac{1}{2}\frac{G_C G_L}{(2G_C^2 + 3G_C G_L + G_L^2)}. \qquad (4.41)$$

The derivative, $\partial \eta / \partial G_c = 0$, shows that

$$R_C = \sqrt{2}R_L, \qquad I_C = \frac{V_{CC}}{2R_L}, \quad \text{and} \quad V_{CE} = \frac{V_{CC}}{2 + \sqrt{2}} \qquad (4.42)$$

for maximum efficiency or minimum quiescent power.

REFERENCES

[1] A. W. Lo et al., *Transistor Electronics*, Prentice-Hall, Englewood Cliffs, N.J., 1955, Chapter 9.
[2] R. F. Shea et al., *Transistor Circuit Engineering*, Wiley, New York, 1957, Chapter 7.
[3] D. Dewitt and A. L. Rossoff, *Transistor Electronics*, McGraw-Hill, New York, 1957, Chapter 10.
[4] J. G. Linvill and J. F. Gibbons, *Transistors and Active Circuits*, McGraw-Hill, New York, 1961, Chapter 19.
[5] M. V. Joyce and K. K. Clarke, *Transistor Circuit Analysis*, Addison-Wesley, Reading, Mass., 1961, Chapter 8.
[6] R. D. Thornton et al., *Multistage Transistor Circuits*, SEEC, Vol. 5, Wiley, New York, 1965, Chapter 5.
[7] G. Brunn, "Common Emitter Transistor Video Amplifiers," *Proc. IRE*, **44**, 1561–1572 (November 1956).
[8] W. E. Ballentine and F. H. Blecher, "Broadband Transistor Video Amplifiers," *Dig. Tech. Papers*, 1959 ISSCC, University of Pennsylvania, Philadelphia, Pa., Vol. II, 42–43 (February 1959).
[9] D. O. Pederson and R. S. Pepper, "An Evaluation of Transistor Lawpass Broadbanding Techniques," 1959 *IRE WESCON Conv. Rec.*, Part 2, **3**, 111–126 (August 1959).
[10] J. W. Baker, "Achieving Stable High-Gain Video Amplification Using the Miller-Effect Transformation," *Proc. IEEE*, **51**, 911–916 (June 1963).
[11] J. D. Meindl and P. H. Hudson, "Low Power Linear Circuits," *IEEE J. Solid-State Circuits*, **1**, 100–111 (December 1966).
[12] J. D. Meindl et al., "Static and Dynamic Performance of Micropower Transistor Linear Amplifiers," in *Micropower Electronics*, E. Keonjian, Ed., Macmillan, New York, 1964.
[13] J. Feit and R. McGinnis, "Small Signal Functional Circuits," USAECOM, Fort Monmouth, N.J., Contract No. DA 28-043 AMC-00151(E), Final Report July 1964–June 1965.
[14] L. J. Sevin, *Field Effect Transistors*, McGraw-Hill, New York, 1965, Chapter 3.

Chapter 5

Tuned Amplifiers

The subject of this chapter is the design of micropower narrow band tuned amplifiers [1–6] for minimum quiescent power. In addition to the noise figure of input stages, the design constraints of tuned amplifiers quite often include gain, bandwidth, center frequency, stability, and selectivity. The primary effects of these constraints on quiescent power are discussed in the following sections.

5.1 THE ITERATIVE STAGE

In this section the gain, center frequency, bandwidth, and stability of a micropower narrow band iterative amplifier stage are considered as a basis for the analysis of the overall amplifier.

Gain

The circuit diagrams of a multistage common emitter tuned amplifier are illustrated in Figure 5.1. Referring to Figure 5.2, with ideal transformer coupling the voltage gain of the iterative stage is

$$A_v = -\frac{V_0}{V_i} \simeq a g_m Z_L$$

$$\simeq \frac{A_{v0}}{1 + j\left\{\omega\left[\dfrac{C_T + a^2 C_{be} + a^2(1 + A_{v0}/a)C_c}{a^2 g_{be}}\right] - \dfrac{1}{\omega}\left[\dfrac{1}{a^2 g_{be} L_T}\right]\right\}}, \quad (5.1)$$

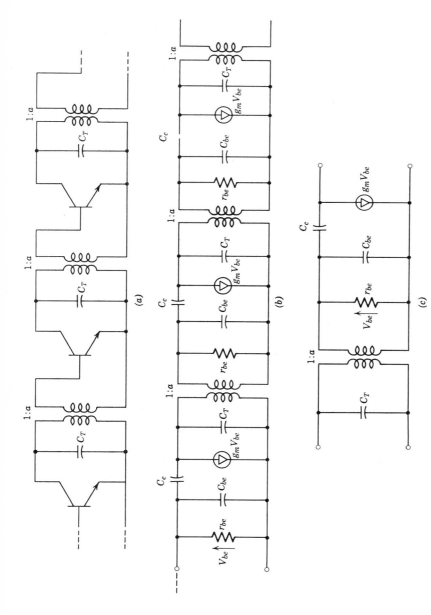

Figure 5.1 Tuned common emitter iterative circuit diagrams: (*a*) ac schematic diagram; (*b*) multistage equivalent circuit; (*c*) iterative stage equivalent circuit.

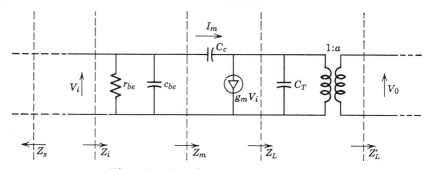

Figure 5.2 Iterative stage equivalent circuit.

assuming $|Z_L| \ll |1/j\omega C_c|$, $|I_m| \ll g_m V_i$, and that *solely* for the purpose of calculating A_v, $1/Z'_L = Y'_L = g_{be} + j\omega[C_{be} + (1 + A_{v0})C_c]$. The bases for the first two of these three assumptions are discussed in connection with (4.5) and (4.6). The third assumption is tantamount to stating that the load of stage $(n + 1)$ is taken as purely real for the purpose of calculating the load of stage n. From (5.1) the voltage gain at the center or resonant frequency is

$$A_v(\omega_0) = A_{v0} = \frac{h_{fe0}}{a}. \tag{5.2}$$

For an iterative stage, the current and voltage gain are equal so that

$$G_{p0} = A_{v0} A_{i0} = A_{v0}^2 = \left(\frac{h_{fe0}}{a}\right)^2 \tag{5.3}$$

is the actual power gain of the iterative stage at the center frequency ω_0. A more general equation for power gain [2],

$$G_p = \frac{|y_{21}|^2 \, Re\, Y_L}{|y_{22} + Y_L|^2 \, Re[y_{11} - y_{12}y_{21}/(y_{22} + Y_L)]}, \tag{5.4}$$

yields the result of (5.3) at ω_0; G_{p0} and A_{v0} can be controlled by choice of the transistor current gain h_{fe0} and the coupling transformer turns ratio a.

Center Frequency

From (5.1) the center or resonant frequency of the iterative stage is given by

$$\omega_0^2 \simeq \left(L_T\left\{C_T + a^2\left[C_{be} + \left(1 + \frac{A_{v0}}{a}\right)C_c\right]\right\}\right)^{-1}. \tag{5.5}$$

The value of ω_0 can be controlled through the tuning capacitance C_T and inductance L_T, which are assumed lossless.

Bandwidth

From (5.1)

$$A_v = \frac{A_{v0}}{1 + jQ(\omega/\omega_0 - \omega_0/\omega)}, \tag{5.6}$$

where ω_0 is given by (5.5) and

$$Q = \omega_0 \frac{C_T + a^2 C_{be} + a^2(1 + A_{v0}/a)C_c}{a^2 g_{be}} = \frac{1}{\omega_0} \frac{1}{a^2 g_{be} L_T}. \tag{5.7}$$

The 3 dB cutoff frequencies of the iterative stage, ω_c, occur when

$$\left|\frac{A_v}{A_{v0}}\right| = \frac{1}{\sqrt{2}} = \left|\left[1 + jQ\left(\frac{\omega_c}{\omega_0} - \frac{\omega_0}{\omega_c}\right)\right]^{-1}\right|, \tag{5.8}$$

which gives the upper and lower cutoff frequencies

$$\omega_c = \frac{1}{2Q}(\sqrt{1 + 4Q^2} \pm 1)\omega_0 \tag{5.9}$$

and their difference

$$\Delta\omega = \frac{\omega_0}{Q} = \frac{a^2 g_{be}}{C_T + a^2 C_{be} + a^2(1 + A_{v0}/a)C_c}, \tag{5.10}$$

the 3 dB bandwidth of the iterative stage. Combining (5.6) and (5.10),

$$A_v = \frac{A_{v0}}{1 + j(\omega/\Delta\omega)[1 - (\omega_0/\omega)^2]} \tag{5.11}$$

illustrates the nearly symmetrical behavior of $|A_v|$ in the vicinity of ω_0.

Stability

The admittance, $Y_m = 1/Z_m$, because of the internal feedback through C_c in the circuit in Figure 5.2, is

$$Y_m = \frac{I_m}{V_i} = \frac{1 + g_m Z_L}{1/j\omega C_c + Z_L} \tag{5.12a}$$

$$\simeq j\omega C_c \frac{A_{v0}/a}{1 + jQ(\omega/\omega_0 - \omega_0/\omega)}, \tag{5.12b}$$

based on the assumptions associated with (5.1) and the substitution of (5.7). A graphical representation of (5.12b) in the vicinity of ω_0 is shown in Figure 5.3a. The possibility of $Re Y_m = G_m < 0$, hence a potentially unstable

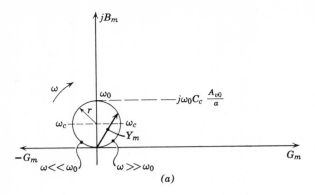

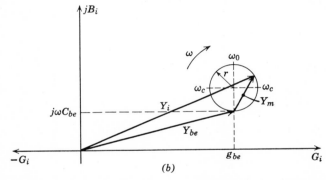

Figure 5.3 Admittance diagrams: (a) reflected load admittance; (b) input admittance.

circuit, is reflected in that portion of the circular locus in Figure 5.3a contained in the second quadrant. In constructing the locus $|\omega C_c(A_{v0}/a)|$ is essentially a constant, whereas the variation in Y_m is dominated by the denominator of (5.12b). The 3 dB cutoff frequencies of the iterative stage, ω_c, are separated by 180° on the locus. The radius of the circle is

$$r = \frac{1}{2}\left|\omega_0 C_c \frac{A_{v0}}{a}\right|. \tag{5.13}$$

From Figure 5.2, the input admittance of the iterative stage, $Y_i = 1/Z_i$, is

$$\begin{aligned}Y_i &= Y_{be} + Y_m \\ &= g_{be} + j\omega C_{be} + Y_m.\end{aligned} \tag{5.14}$$

Assuming that the magnitude of the source conductance is small compared with that of the input conductance or

$$|Re\,Y_s| = |G_s| \ll |Re\,Y_i|, \tag{5.15}$$

the iterative stage will be potentially unstable unless $Re\,Y_i > 0$. From Figure 5.3b it is evident that if the radius of the circle, r, which describes the locus of Y_m, is smaller than the coordinate of its center on the G_i axis, g_{be}, then $Re\,Y_i > 0$ and the iterative stage is unconditionally stable. That is,

or
$$K = \frac{r}{g_{be}} < 1$$

$$K = \frac{1}{2}\frac{\omega_0 C_c}{g_m}\left(\frac{h_{fe0}}{a}\right)^2 < 1 \tag{5.16a}$$

provides assurance that the iterative stage will be stable regardless of the passive element terminations of the stage. The general equation for the stability factor [2,4,5,7,8,9],

$$K = \frac{|y_{12}y_{21}|}{2(g_{11} + G_s)(g_{22} + G_L) - Re(y_{12}y_{21})}, \tag{5.16b}$$

yields the result of (5.16a) for the micropower iterative stage. It is important to recognize that $(g_{11} + G_s) > 0$ and $(g_{22} + G_L) > 0$ are assumed in the derivation of (5.16a) and (5.16b).

From the preceding discussion, it is evident that $K = r/g_{be} < 1$ is required for unconditional stability and the value of K can be taken as a measure of the stability of the micropower stage. In addition, from (5.14) and Figure 5.3b, it is evident that the condition $Y_m \ll Y_{be}$ ensures a small interaction between stages in the amplifier. This permits convenient and accurate alignment and prevents skewing of the bandpass characteristic.

5.2 THE COMMON EMITTER AMPLIFIER

The overall amplification of a micropower tuned amplifier can be expressed as

$$A_v^* = A_v^{\,n} = \left[\frac{A_{v0}}{1 + jQ(\omega/\omega_0 - \omega_0/\omega)}\right]^n \tag{5.17}$$

on the basis of (5.6). Consequently, the overall amplification and power gain at the center frequency ω_0 are

$$A_{v0}^* = A_{v0}^{\,n} \quad \text{and} \quad G_{p0}^* = G_{p0}^{\,n} = A_{v0}^{\,2n}, \tag{5.18}$$

respectively. For $K \ll 1$, each stage, and consequently the overall amplifier, has a center frequency ω_0. From (5.17), the overall 3 dB cutoff frequencies of the amplifier, ω_c^*, occur when

$$\left|\frac{A_v^*}{A_v}\right|^2 = (\tfrac{1}{2}) = \left[1 + Q^2\left(\frac{\omega_c^*}{\omega_0} - \frac{\omega_0}{\omega_c^*}\right)^2\right]^{-n} \tag{5.19}$$

and their difference

$$(\Delta\omega)^* = \sqrt{2^{1/n} - 1}\,\frac{\omega_0}{Q} = \sqrt{2^{1/n} - 1}\,\Delta\omega \tag{5.20a}$$

$$\simeq \frac{\Delta\omega}{1.2\sqrt{n}} \tag{5.20b}$$

is the overall bandwidth of the amplifier.

The quiescent power of the amplifier is

$$P_{DC}^* = nV_{CC}I_C. \tag{5.21}$$

Substituting (5.3) and (5.15) into (5.21) gives

$$P_{DC}^* = \frac{0.5\gamma V_{CC}\omega_0 C_c}{K} n(G_{p0}^*)^{1/n} \tag{5.22a}$$

or

$$P_{DC}' = \frac{P_{DC}^*}{0.5\gamma V_{CC}\omega_0 C_c(1/K)} = n(G_{p0}^*)^{1/n} \tag{5.22b}$$

in normalized form [10]. This result indicates the following for minimum power drain P_{DC}^* in a micropower tuned amplifier:

1. The internal transistor feedback capacitance C_c should be as small as possible. This, of course, applies as well to stray feedback capacitance in the circuit.
2. The center frequency ω_0, if flexible, should be chosen as small as possible.
3. The stability factor K should not be excessively small since a quiescent power-stability factor trade-off applies.
4. The number of stages n in the amplifier should be carefully chosen.

Figure 5.4 displays a normalized plot of (5.22b), which indicates that by increasing the number of stages of a tuned iterative amplifier beyond what would be considered the minimum number for a fixed G_{p0}^*, a very substantial quiescent power savings is possible.

From (5.22), the optimum value for n can be found by taking the derivative $\partial P_{DC}'/\partial n$, which yields

$$G_{p0} = \epsilon = 2.72 = 4.35 \text{ dB} \quad \text{and} \quad n = \ln G_{p0}^* \tag{5.23}$$

as the optimum gain per stage and number of stages, respectively, for minimum P_{DC}^* [10].

A graphical summary of the more prominent features of a micropower tuned iterative stage is provided by Figure 5.5. The design begins with the

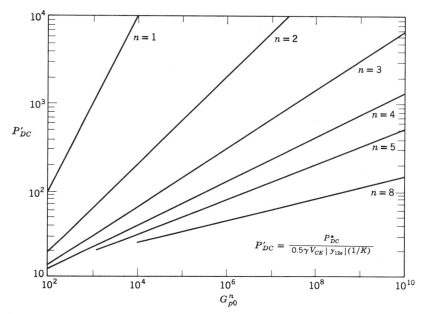

Figure 5.4 Normalized quiescent power P'_{DC} for a tuned common emitter iterative amplifier versus overall power gain at the center frequency for various numbers of stages, n.

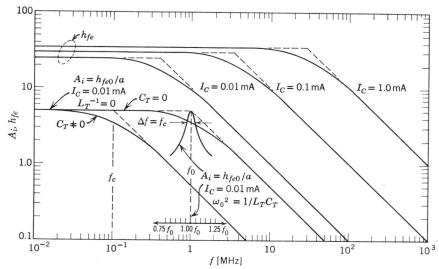

Figure 5.5 Transistor current gain with $I_C = 1.0, 0.1, 0.01$ mA, wideband amplifier current gain with $C_T = 0$ and $C_T = 9[C_{be} + (1 + A_{v0}/a)C_c] \neq 0$, and narrow band amplifier current gain with $L_T = C_T\omega_0^2$ versus frequency.

selection of a transistor whose maximum gain-bandwidth product is well beyond the range of direct interest for the circuit under consideration (e.g., $f_T = 1000$ MHz at $I_C = 1.0$ mA). The quiescent collector current of the transistor is then reduced until K, the stability factor, (5.15), increases to the specified design value. This provides a substantial reduction in quiescent current (e.g., $I_C = 0.01$ mA). The transformer turns ratio, a, is then adjusted to give the required current gain (e.g., $A_i = h_{feo}/a = 5$). The tuning capacitance C_T is then added to the circuit to reduce its bandwidth from f_0 to the design value $\Delta f = f_c$. Finally, the tuning inductance L_T is added to provide selective amplification in a narrow band around the resonant frequency f_0.

5.3 THE CASCODE AMPLIFIER

The circuit diagrams of a tuned cascode iterative amplifier [10,11,12] are illustrated in Figure 5.6. The power gain of the iterative stage is described by (5.4) if the overall fourpole parameters of the cascode stage are used. Since

$$|Y| = \begin{vmatrix} \dfrac{y_{11e}(y_{22e} + y_{11b}) - y_{12e}y_{21e}}{y_{22e} + y_{11b}} & -\left(\dfrac{y_{12e}y_{12b}}{y_{22e} + y_{11b}} + y_{12s}\right) \\ \dfrac{-y_{21e}y_{21b}}{y_{22e} + y_{11b}} & \dfrac{y_{22b}(y_{22e} + y_{11b}) - y_{12b}y_{21b}}{y_{22e} + y_{11b}} \end{vmatrix} \quad (5.24a)$$

$$\simeq \begin{vmatrix} y_{11e} & -\left(\dfrac{y_{12e}y_{12b}}{y_{11b}} + y_{12s}\right) \\ y_{21e} & h_{22b} \end{vmatrix} \quad (5.24b)$$

describe the cascode stage, (5.4) gives

$$G_{p0} \simeq \left(\dfrac{h_{feo}}{a}\right)^2 \quad (5.25)$$

as the power gain at the center frequency ω_0. In a micropower cascode stage, often the overall reverse transadmittance of the stage due to parasitic circuit elements, y_{12s}, is larger than the overall reverse transadmittance of the stage due to internal transistor feedback, $y_{12e}y_{12b}/y_{11b}$. Parasitic circuit capacitance between points x and x' of Figure 5.6a is a principal cause of y_{12s}. The total overall reverse transadmittance of a cascode stage is $-[(y_{12e}y_{12b}/y_{11b}) + y_{12s}]$.

From (5.16) and (5.24b), the stability factor is

$$K \simeq \dfrac{1}{2}\left|\dfrac{y_{12e}y_{12b}}{y_{11b}} + y_{12s}\right|\dfrac{G_{p0}}{g_m}. \quad (5.26)$$

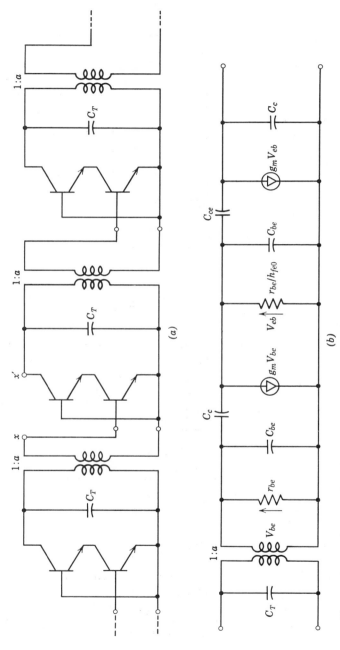

Figure 5.6 Tuned cascode iterative amplifier circuit diagrams: (*a*) ac schematic diagram; (*b*) iterative stage equivalent circuit.

94   **Tuned Amplifiers**

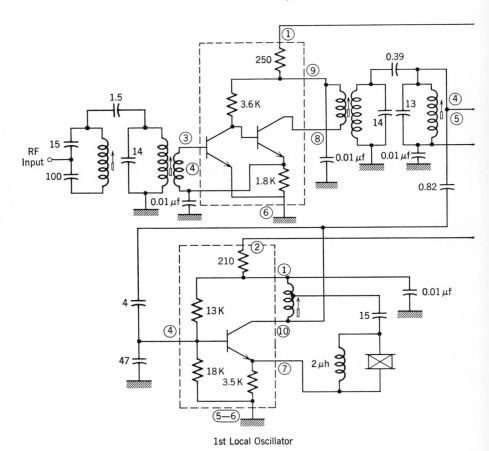

Figure 5.7 Schematic diagrams of a micropower *RF* tuned

Consequently the total quiescent power is

$$P_{DC}^* = nV_{CC}I_C$$
$$= 0.5\gamma V_{CC} \left| \frac{y_{12e}y_{12b}}{y_{11b}} + y_{12s} \right| \frac{1}{K} n(G_{p0}^*)^{1/n}, \tag{5.27}$$

which is comparable to (5.22a) for the common emitter iterative amplifier. A direct comparison of (5.22a) and (5.27) indicates

$$\frac{P_{DC}^*(\text{CE})}{P_{DC}^*(\text{cascode})} \simeq \frac{|y_{12e}|}{|y_{12e}y_{12b}/y_{11b} + y_{12s}|} \tag{5.28}$$

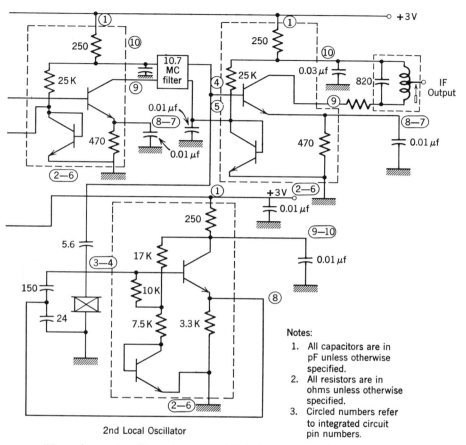

amplifier, mixer and oscillator (courtesy of U.S. Army Electronics Command [13]).

for equal gain, stability factor, and number of stages for the two amplifiers. From (5.28) it is evident that the cascode amplifier in Figure 5.6 can provide a savings in quiescent power compared to the common emitter amplifier if $|y_{12e}| > |y_{12e}y_{12b}/y_{11b} + y_{12s}|$, which is likely to occur in a micropower circuit.

From (5.27), $\partial P_{DC}^*/\partial n$ indicates that

$$G_{p0} = \epsilon = 2.72 = 4.35 \text{ dB} \quad \text{and} \quad n = \ln G_{p0}^* \quad (5.29)$$

are the optimum gain per stage and number of stages, respectively, for minimum P_{DC}^* in a cascode amplifier.

96 Tuned Amplifiers

In the tuned common emitter iterative amplifier, the tendency toward instability resulting from internal transistor feedback was controlled by operating the stage with a large load conductance to prevent oscillation. In the cascode stage the tendency toward instability is counteracted further by reducing the internal feedback. A third method for combating instability and reducing quiescent power is to employ neutralization [1–5]. Where the effective feedback transconductance Y_{12} of a stage becomes extremely small, the minimum quiescent power is influenced by the finite Q of the unloaded tuned circuit.

The selectivity of a tuned amplifier depends primarily on passive interstage coupling networks [7] and consequently does not ordinarily influence P_{DC}^* significantly.

5.4 A MONOLITHIC TUNED AMPLIFIER

Figure 5.7 illustrates the schematic diagram of a micropower RF tuned amplifier utilizing a monolithic integrated circuit (enclosed within the dashed

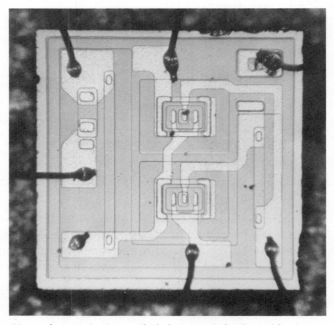

Figure 5.8 Photomicrograph of monolithic integrated circuit used in the tuned amplifier of Figure 5.7. Emitter strip width is 1 mil, chip size is 30 × 30 mils (courtesy of U.S. Army Electronics Command [13]).

rectangle) as well as several other circuit diagrams. Figure 5.8 shows a photomicrograph of the monolithic chip of the RF amplifier. The salient performance characteristics of the circuit are listed [13]:

power gain	16–18 dB
noise figure	5–7 dB
operating frequency	51 MHz
3 dB bandwidth	1 MHz
nominal supply voltage	3.0 V
power drain	2.4 mW

The maximum gain-bandwidth product of the transistors used in the circuit is $f_T \simeq 450$ MHz at $I_C = 5.0$ ma, $V_{CE} = 5.0$ V. By operating at $I_C = 400$ μa with $f_T \simeq 140$ MHz and using two common emitter transistors in cascade (rather than a single device), adequate power gain is maintained, noise figure is slightly enhanced, internal feedback is reduced (thus permitting improved RF selectivity), and power drain is markedly decreased.

REFERENCES

[1] R. F. Shea et al., *Transistor Circuit Engineering*, Wiley, New York, 1957, Chapter 6.
[2] J. G. Linvill and J. F. Gibbons, *Transistors and Active Circuits*, McGraw-Hill, New York, 1961, Chapters 11 and 18.
[3] M. V. Joyce and K. K. Clarke, *Transistor Circuit Analysis*, Addison-Wesley, Reading, Mass., 1961, Chapter 9.
[4] C. L. Searle et al., *Elementary Circuit Properties of Transistors*, SEEC, Vol. 3, Wiley, New York, 1964, Chapter 8.
[5] R. D. Thornton et al., *SEEC*, Vol. 5, *Multistage Transistor Circuits*, Wiley, New York, 1965, Chapter 7.
[6] J. M. Petit and M. M. McWhorter, *Electronic Amplifier Circuits*, McGraw-Hill, New York, Chapter 7, 1961.
[7] A. P. Stern, "Stability and Power Gain of Tuned Transistor Amplifiers," *Proc. IRE*, **45**, 335–343 (March 1977).
[8] E. F. Bolinder, "Survey of Some Properties of Linear Networks," *IRE Trans. Circuit Theory*, **4**, 70–78 (September 1957).
[9] J. M. Rollett, "Stability and Power Gain Invariants of Linear Two-ports," *IRE Trans. Circuit Theory*, **9**, 29–32 (March 1962).
[10] J. D. Meindl and P. H. Hudson, "Low Power Linear Circuits," *IEEE J. Solid-State Circuits*, **1**, 100–111 (December 1966).
[11] J. R. James, "Analysis of the Cascode Transistor Configuration," *Electron. Eng.*, **32**, 44–48 (January 1960).
[12] Y. G. Kryakov and Y. L. Simonov, "Analysis of a Cascode Transistor Tuned Amplifier," *Electric Commun.*, **37**, 40–44 (January 1962).
[13] J. Feit and R. McGinnis, "Small Signal Functional Circuit Units," USAECOM, Fort Monmouth, N.J., Contract No. DA 28-043 AMC-00151(E), Final Report, July 1964–June 1965.

Chapter 6

Low-Noise Amplifiers

Since it largely determines the minimum detectable signal of an amplifier, the noise figure of the input stage is often its primary design constraint. Secondary specifications for the input stage frequently include gain, bandwidth, center frequency, stability, and other requirements. The purpose of this chapter is to discuss the design of low-noise input stages for minimum quiescent power [1–8].

6.1 A TRANSISTOR NOISE MODEL

The noise equivalent circuit of a junction transistor is illustrated in Figure 6.1. Five noise generators have been added to the hybrid pi model in Figure 1.6. For a narrow frequency interval Δf, the flicker noise generator

$$\overline{i_f^2} = K_f I_B^\lambda \frac{1}{f^\alpha} \Delta f \tag{6.1}$$

is characterized by its $1/f$ frequency dependence and cannot be predicted accurately. Typically it will dominate at frequencies below the region of 1000 Hz. In (6.1) K_f, λ, and α must be determined from measurement. Both λ and α are often about unity, whereas K_f varies greatly from one transistor to another. Surface generation or recombination of carriers and temperature fluctuations are causes of flicker noise.

The thermal noise caused by random motion of carriers in the extrinsic

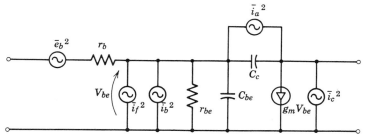

Figure 6.1 Hybrid pi noise model for the transistor.

base resistance r_b is represented by the generator

$$\overline{e_b^2} = 4kTr_b\,\Delta f \qquad (6.2)$$

in the frequency interval Δf.

The mean square values of the three-shot noise generators in a narrow frequency interval are

$$\overline{i_b^2} = 2qI_B\,\Delta f, \qquad (6.3a)$$

$$\overline{i_c^2} = 2qI_C\,\Delta f, \qquad (6.3b)$$

and

$$\overline{i_a^2} = 2qI_{CB0}\,\Delta f. \qquad (6.3c)$$

The random recombination of carriers in the base region that contributes to fluctuations in the dc current I_B is represented by $\overline{i_b^2}$. The random arrival of minority carriers by diffusion at the collector junction contributes to fluctuations in I_C and is represented by $\overline{i_c^2}$. The collector junction reverse current I_{CB0} is produced by random thermal generation of carriers and the fluctuation of I_{CB0} is represented by $\overline{i_a^2}$. The latter generator is negligible in a good low-noise transistor because of the small value of I_{CB0}. The noise sources described by (6.1), (6.2), and (6.3) are assumed uncorrelated.

6.2 NOISE FIGURE

The effect at the output of the amplifier illustrated in Figure 6.2 of all of the noise sources in the transistor can be represented by a single equivalent noise generator $\overline{e_{eq}^2}$ located in series with the source noise generator $\overline{e_s^2}$. In a narrow frequency interval

$$\overline{e_{eq}^2} = \overline{e_b^2} + \overline{i_b^2}(R_s + r_b)^2 + \overline{i_f^2}(R_s + r_b)^2 + \overline{i_c^2}\frac{|R_s + r_b + Z_{be}|^2}{|g_m Z_{be}|^2}, \qquad (6.4)$$

assuming

$$|Z_{be}| = \left|\frac{r_{be}}{1 + j\omega r_{be} C_{be}}\right| \ll \left|\frac{1}{j\omega C_c}\right|,$$

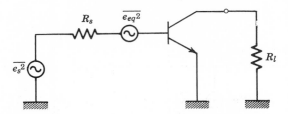

Figure 6.2 AC circuit diagram of a low noise amplifier stage illustrating the noise sources.

can be derived from the noise model in Figure 6.1. The equivalent noise resistance of the amplifier is

$$R_{eq} = \frac{\overline{e_{eq}^2}}{4kT_0 \Delta f}, \tag{6.5}$$

where $T_0 = 290°K$. The noise figure F of the amplifier in Figure 6.2 is defined as the ratio of the total noise power delivered by the amplifier to R_l to the noise power that would be delivered to R_l if the only noisy component were the source resistance R_s at a temperature T_0. Consequently

$$F = \frac{\overline{e_{eq}^2} + \overline{e_s^2}}{\overline{e_s^2}} = \frac{R_{eq} + R_s}{R_s} \tag{6.6}$$

is independent of R_l. Substituting (6.1), (6.2), and (6.3) into (6.6) gives

$$F = 1 + \frac{r_b}{R_s} + \frac{K_f}{4kT_0} I_B^\lambda \frac{(R_s + r_b)^2}{R_s} f^{-\alpha} + \frac{1}{2\gamma} \frac{I_C}{h_{FE}} \frac{(R_s + r_b)^2}{R_s}$$
$$+ \frac{1}{2\gamma} \frac{I_C}{h_{feo}^2} \frac{(R_s + r_b + r_{be})^2}{R_s} + \frac{1}{2\gamma} I_C \frac{(R_s + r_b)^2}{R_s} \left(\frac{f}{f_T}\right)^2. \tag{6.7}$$

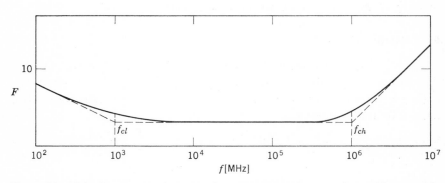

Figure 6.3 Noise figure versus frequency for a common emitter transistor with R_s and I_C constant.

From this result (Figure 6.3), it is evident that at very low frequencies F increases as $1/f$ or at a rate of about 3 dB per octave (for $\alpha = 1$). At high frequencies F increases as f^2 or at the rate of 6 dB per octave. Normally, F reaches a constant minimum value in a midband range extending from the lower-noise corner frequency f_{cl} to the upper noise corner frequency f_{ch}. It is assumed that all passive circuit elements, such as coupling capacitors and biasing resistors, constitute a negligible influence on the noise behavior of the circuit.

6.3 LOW-FREQUENCY AMPLIFIERS

The design of low-noise amplifiers operating in a low-frequency range in which the final term in (6.7) is negligible is considered in this section. The effects of several common design constraints on the quiescent power are discussed.

Flicker Noise Region

For the low-frequency range in which flicker noise dominates, (6.7) gives

$$F_f = 1 + \frac{K_f}{4kT_0} I_B^\lambda \frac{(R_s + r_b)^2}{R_s} \frac{1}{f^\alpha}. \tag{6.8}$$

For a fixed operating point, $\partial F_f / \partial R_s$ indicates that the minimum flicker noise figure [8,9]

$$F_f(\min) = 1 + \frac{K_f}{kT_0} I_B^\lambda r_b \frac{1}{f^\alpha} \tag{6.9}$$

occurs when the effective value of R_s is adjusted so that $R_s = r_b$. This condition actually provides a convenient means of measuring r_b [9].

Shot Noise Region

For the midband range in which shot noise dominates, (6.7) gives

$$F_0 = 1 + \frac{r_b}{R_s} + \frac{1}{2\gamma} \frac{I_C}{h_{FE}} \frac{(R_s + r_b)^2}{R_s} + \frac{1}{2\gamma} \frac{I_C}{h_{fe0}^2} \frac{(R_s + r_b + r_{be})^2}{R_s}. \tag{6.10}$$

If the dc operating point of the transistor is fixed (for example, by the required value of the upper noise corner frequency), $\partial F_0 / \partial R_s$ indicates that a minimum value of (6.10)

$$F_0(\min)\bigg|_{I_C = \text{constant}} = 1 + \left(\frac{1 + 2g_m r_b}{h_{FE}}\right)^{1/2} \tag{6.11}$$

occurs for an optimum value of source resistance given by

$$R_s^2 = r_b^2 + h_{FE}\frac{1 + 2g_m r_b}{g_m^2} \simeq h_{FE}\frac{1 + 2g_m r_b}{g_m^2}. \qquad (6.12)$$

From (6.11) it is evident that a small base spreading resistance r_b and a large value of current gain h_{FE} are desirable characteristics of a low-noise transistor.

If transformer coupling is undesirable and the source resistance is fixed but I_C is flexible, $\partial F_0/\partial I_C$ indicates that a minimum value of (6.10)

$$F_0(\min)\bigg|_{R_s=\text{constant}} = 1 + \frac{r_b}{R_s} + \frac{1}{\sqrt{h_{FE}}}\frac{R_s + r_b}{R_s} \qquad (6.13)$$

occurs for an optimum value of collector current given by

$$I_C^2 = \gamma^2 \frac{h_{FE}}{(R_s + r_b)^2}, \qquad (6.14)$$

assuming that $\partial h_{FE}/\partial I_C = 0$. From (6.13), it is evident that $R_S \gg r_b$ is a desirable condition for a small F_0.

If the effective source resistance as well as the operating point is adjustable, a true minimum for the shot noise figure can be calculated [5]. Since $\partial F_0/\partial R_S = 0$ and $\partial F_0/\partial I_C = 0$ are the conditions for this minimum, $\partial F_0/\partial R_S$ gives, from (6.11),

$$F_0 = 1 + \left(\frac{1 + 2g_m r_b}{h_{FE}}\right)^{1/2}. \qquad (6.15)$$

From (6.15) $\partial F_0/\partial I_C$ gives

$$\frac{2g_m r_b}{1 + 2g_m r_b} = I_B \frac{\partial h_{FE}}{\partial I_C}, \qquad (6.16)$$

assuming r_b is independent of I_C. From (1.13), if I_B is dominated by excess current components so that

$$I_C = \alpha_N I_{EBS} \epsilon^{qV_{BE}/kT} \qquad (6.17a)$$

and

$$I_B = I_{EX} \epsilon^{qV_{BE}/nkT}, \qquad (6.17b)$$

(1.14) gives

$$h_{FE} = \frac{I_C}{I_B} = \frac{(\alpha_N I_{EBS})^{1/n}}{I_{EX}} I_C^{(n-1)/n} \qquad (6.18)$$

and

$$\frac{\partial h_{FE}}{\partial I_C} = \frac{1}{I_B}\frac{n-1}{n}, \qquad (6.19)$$

where $n = h_{feo}/h_{FE} \geq 1$ is the emitter junction ideality factor. Combining (6.16) and (6.19),

$$I_C = \tfrac{1}{2}\gamma \frac{n-1}{r_b} \tag{6.20}$$

gives the optimum quiescent collector current for minimum F_0. Substituting this result into (6.12) yields

$$R_s^2 = r_b^2 + h_{FE} \frac{4n}{(n-1)^2} r_b^2, \tag{6.21}$$

which describes the optimum source resistance for minimum F_0. Finally, (6.20) and (6.11) give

$$F_0(\min) = 1 + \left(\frac{n}{h_{FE}}\right)^{1/2}, \tag{6.22}$$

the minimum noise figure in the shot noise region for optimum source resistance and operating point.

As $n \to 1$, h_{FE} tends to become independent of I_C and (6.20) indicates $I_C \to 0$ as a condition for minimum F_0. On the basis of (6.20)–(6.22) the importance of excess emitter junction current (i.e., n) and base spreading resistance r_b in determining the optimum values of I_C and R_s for minimum F_0 is apparent.

The quiescent power of a low-frequency low-noise amplifier stage of optimum design can be expressed as

$$P_{DC}^* = V_{CC}I_C = \tfrac{1}{2}\gamma V_{CC}\frac{n-1}{r_b}, \tag{6.23}$$

which for the typical conditions $\gamma = 0.026$ V, $V_{CC} = 3.0$ V, $n = 1.5$ and $r_b = 100\ \Omega$ yields a quiescent power $P_{DC}^* = 195\ \mu$W at a collector current of 65 μA. Obviously, reducing emitter junction current or n improves both F_0 and P_{DC}^*. A plot of (6.10) in the form of contours of constant noise figure [5,8,9], F, is illustrated in Figure 6.4. The point described by (6.20)–(6.22) lies within the $F = 1.1$ contour.

6.4 HIGH-FREQUENCY AMPLIFIERS

For a high-frequency low-noise amplifier stage, the final term in (6.7) becomes important. Assuming the source is transformer coupled to the input stage, an optimum effective value for R_s described by (6.12) can be achieved. For this condition (6.7) reduces to

$$F = 1 + \left(\frac{1 + 2g_m r_b}{h_{FE}}\right)^{1/2}\left[1 + \frac{h_{FE}}{2}\left(\frac{f}{f_T}\right)^2\right]. \tag{6.24}$$

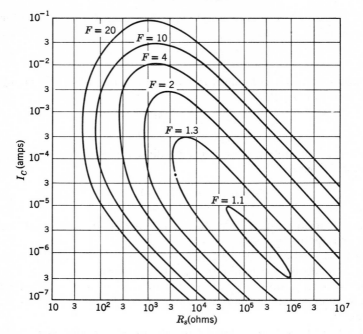

Figure 6.4 Contours of constant low-frequency noise figure for a typical low-noise transistor [8].

From (6.24) the upper noise corner frequency of the amplifier stage, f_{ch}, which occurs when $F - 1$ has increased 3 dB above its midband value, is given by

$$f_{ch}^2 = \frac{2}{h_{FE}} f_T^2 \qquad (6.25a)$$

or

$$\omega_{ch}^2 = \frac{2}{h_{FE}} \left(\frac{g_m}{C_{be} + C_c} \right)^2. \qquad (6.25b)$$

The collector current must be sufficiently large so that the upper noise corner frequency ω_{ch} and the maximum signal frequency ω_s of the stage coincide [10]. This requirement determines the minimum quiescent power of the low-noise high-frequency stage.

$$P_{DC}^* = V_{CC} I_C = \gamma V_{CC} \left(\frac{h_{FE}}{2} \right)^{1/2} (C_{be} + C_c) \omega_s. \qquad (6.26)$$

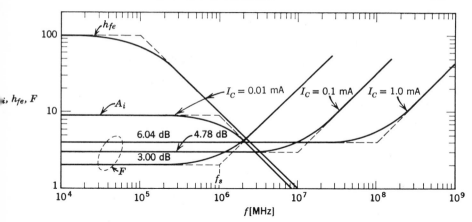

Figure 6.5 Amplifier noise figure F (with $I_C = 1.0$, 0.1, and 0.01 mA), transistor current gain h_{fe}, and amplifier current gain A_i (with $I_C = 0.01$ mA) versus frequency.

By selecting a transistor with small capacitances, C_{be} and C_c, the value of P_{DC}^* can be minimized.

A graphical summary of the more prominent features of the design of a micropower high-frequency low-noise amplifier stage is provided by Figure 6.5. The design begins with the selection of a low-noise transistor whose maximum upper noise corner frequency is well beyond the range of direct interest for the circuit under consideration (e.g., $f_{ch} \simeq 100$ MHz at $I_C = 1.0$ mA). The quiescent collector current is then reduced until the upper noise corner frequency coincides with the maximum signal frequency (e.g., $f_s = 1$ MHz). This provides a substantial reduction in quiescent current (e.g., $I_C = 10$ μA) and for optimum source resistance a current amplification

$$A_i = -\frac{I_l}{I_s} \simeq -\frac{g_m V_{be}}{V_{be}/R_s} \simeq -\sqrt{h_{FE}}, \qquad (6.27)$$

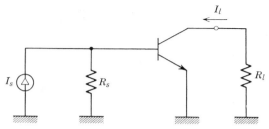

Figure 6.6 AC circuit diagram of a low-noise amplifier stage illustrating source and load signal current.

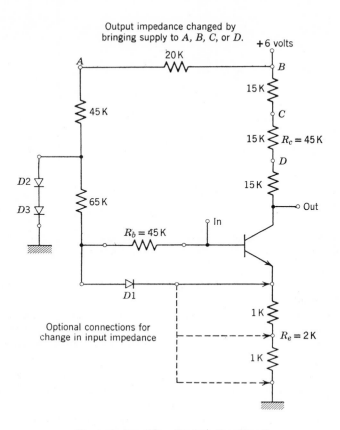

Channel Amplifier Design Specifications

Amplifier Parameters	Design Values	Typical Values
Input impedance (min)	50,000 ohms	80,000 ohms
Output impedance (max)	15,000 ohms	11,000 ohms
Noise figure (R_s = 10,000 ohms), (max)	5.0 dB	4.0 dB
Voltage gain (R_L = 47,000 ohms)	93 ± 1 dB	93 ± 1 dB
Power dissipation (max)	5 mW	4.0 mW
Center frequency (nom)	4.0 kHz	4.0 kHz
Bandwidth (R_s = 10,000 ohms), (max)	7.5 kHz	4.0 kHz
Operating temperature (min)	−20 to +70°C	−40 to +80°C
Supply voltage (max)	+6 V dc	+6 V dc
Package size	⅜ × ¾ × 0.5 in.	same

Figure 6.7 Schematic diagram of iterative stage of a five-stage monolithic micropower low-noise amplifier (after Hoffman [11]).

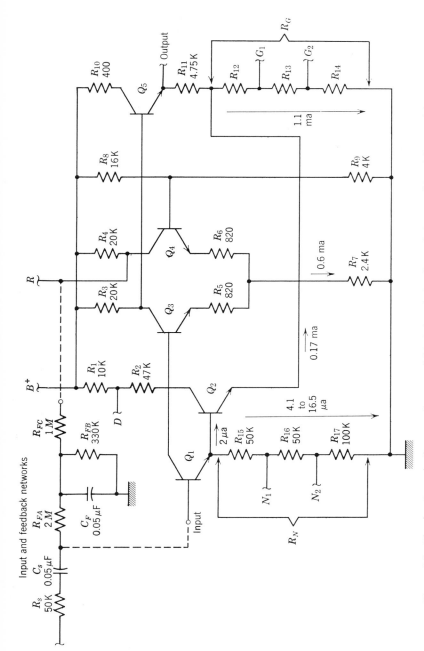

Figure 6.8 Circuit schematic diagram of a monolithic infrared signal translating amplifier (after Maticich [12]).

where I_s is the source signal current and I_l the load signal current as illustrated in Figure 6.6.

6.5 MONOLITHIC LOW-NOISE AMPLIFIERS

The schematic diagram of the iterative stage of a five-stage micropower low-noise amplifier fabricated as a monolithic integrated circuit [11] is illustrated in Figure 6.7a. The overall performance of this capacitor coupled audio amplifier is described below. The circuit is used as a channel amplifier for an infrared mosaic detector.

A second micropower, low-noise, monolithic, infrared signal amplifier [12] is described in Figure 6.8. The input and feedback networks are discrete components while the forward gain is provided by a direct coupled monolithic integrated circuit. Open-loop gain is typically on the order of 1500 and the maximum closed-loop gain used is 90. The supply voltage is approximately 11.5 V.

The schematic diagram of a micropower low-noise transducer amplifier [13] is illustrated in Figure 6.9. In order to maintain a high-input impedance in this monolithic integrated circuit without using impractically large resistors, a novel biasing scheme is used. The gate-bias voltage of the input junction gate field effect transistor is supplied through a diode whose forward bias current is equal to the reverse current of the gate. Consequently, the incremental resistance of the diode is extremely large (for example, $25 \times 10^6 \, \Omega$).

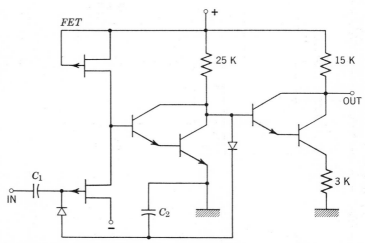

Figure 6.9 Monolithic low-noise transducer amplifier schematic diagram (after Lin [13]).

The salient performance characteristics of the amplifier are listed:

input impedance (400 Hz)	$5 \times 10^6\ \Omega$
output impedance	$10^4\ \Omega$
minimum output voltage	1 volt rms
lower cutoff frequency	300 Hz
upper cutoff frequency	30,000 Hz
minimum voltage gain	250
maximum power dissipation	5 mW
noise figure ($2.5 \times 10^6\ \Omega$ source resistance)	3 dB
supply voltages	$+6, -3$ V

REFERENCES

[1] A. Van der Ziel, *Noise*, Prentice-Hall, Englewood Cliffs, N.J., 1954.
[2] J. G. Linvill and J. F. Gibbons, *Transistors and Active Circuits*, McGraw-Hill, New York, 1961, Chapter 17.
[3] M. V. Joyce and K. K. Clarke, *Transistor Circuit Analysis*, Addison-Wesley, Reading, Mass., 1961, Chapter 7.
[4] D. Dewitt and A. L. Rossoff, *Transistor Electronics*, McGraw-Hill, New York, 1957, Chapter 16.
[5] R. D. Thornton et al., *Characteristics and Limitations of Transistors*, SEEC, Vol. 4, Wiley, New York, 1966, Chapter 4.
[6] E. G. Nielson, "Behavior of Noise Figure in Junction Transistor," *Proc. IRE*, **45**, 957–963 (July 1957).
[7] A. Van der Ziel, "Noise in Junction Transistors," *Proc. IRE*, **46**, 1019–1038 (June 1958).
[8] E. R. Chenette, "Low Noise Transistor Amplifiers," *Solid State Design*, **5**, 27–30 (February 1964).
[9] J. F. Gibbons, "Low-Frequency Noise and Its Application to the Measurement of Certain Transistor Parameters," *IRE Trans. Electron. Devices*, **9**, 308–315 (May 1962).
[10] J. D. Meindl and P. H. Hudson, "Low Power Linear Circuits," *IEEE J. Solid-State Circuits*, **1**, 100–111 (December 1966).
[11] C. P. Hoffman and M. N. Guiliano, "Micropower Molecular Amplifiers for Infrared Search-Track System Application," *Solid-State Design*, **6**, 24–27 (March 1965).
[12] J. R. Maticich, "Design Considerations for an Integrated Low Noise Preamplifier," *Proc. IEEE*, **53**, 605–614 (June 1965).
[13] H. C. Lin and E. A. Karcher, "A Low Noise Integrated Unibi Amplifier with Novel Biasing Scheme and Structure," *Dig. Tech. Papers of 1965 ISSCC*, Philadelphia, Pennsylvania, Vol. XIII 114–115 (February 1965).

Chapter 7

Mixers and Detectors

Two of the more important types of micropower quasilinear circuits are mixers and detectors. The purpose of this chapter is to illustrate the design of these circuits for minimum quiescent power subject to representative performance constraints.

7.1 MIXERS

Typical design constraints for a transistor mixer [1,2,3] may include conversion gain, signal frequency ω_s, local oscillator frequency ω_l, intermediate or output frequency ω_0, and noise figure as well as other requirements. Frequently, the noise performance of a mixer is of paramount importance in determining the sensitivity of a receiver.

The Equivalent Circuit of a Mixer

The simplified circuit diagrams [2,3] of a transistor mixer are indicated in Figure 7.1. The principle of operation of the circuit consists of modulating the quiescent collector current of the transistor and, therefore, its small signal equivalent circuit elements, by means of the local oscillator injection voltage $v_l(t)$ to produce an output frequency $\omega_0 = \omega_l - \omega_s$.

From (1.13), the transistor quiescent collector current can be written as

$$I_C = \alpha_N I_{EBS} \epsilon^{qV_{BE}/kT} = \alpha_N I_{EBS} \epsilon^{V_{BE}/\gamma}. \tag{7.1}$$

If the instantaneous local oscillator voltage appearing across the input

Mixers

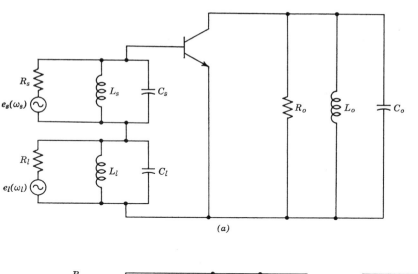

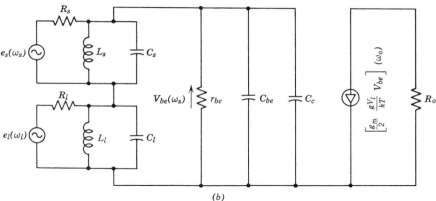

Figure 7.1 Simplified circuit diagrams for a transistor mixer: (*a*) ac schematic diagram; (*b*) simplified equivalent circuit.

terminals of the transistor is

$$v_l(t) = V_l \sin \omega_l t, \tag{7.2}$$

(7.1) can be rewritten as

$$i_C(t) = \alpha_N I_{EBS} \epsilon^{V_{BE}/\gamma} \epsilon^{v_l/\gamma}$$

$$= I_{CQ}\left[1 + \frac{v_l}{\gamma} + \frac{1}{2!}\left(\frac{v_l}{\gamma}\right)^2 + \frac{1}{3!}\left(\frac{v_l}{\gamma}\right)^3 + \cdots\right] \tag{7.3}$$

where $I_{CQ} = \alpha_N I_{EBS} \epsilon^{v_l/\gamma}$.

Substituting (7.2) into (7.3) gives

$$
\begin{aligned}
i_C(t) = I_{CQ}\Bigg\{ &\left[1 + \frac{1}{4}\left(\frac{V_l}{\gamma}\right)^2 + \frac{1}{4}\cdot\frac{1}{16}\left(\frac{V_l}{\gamma}\right)^2 + \cdots \right] \\
&+ \left[\frac{V_l}{\gamma} + \frac{1}{2\cdot 4}\left(\frac{V_l}{\gamma}\right)^3 + \frac{1}{3\cdot 4\cdot 16}\left(\frac{V_l}{\gamma}\right)^5 + \cdots \right]\sin \omega_l t \\
&+ \left[-\frac{1}{4}\left(\frac{V_l}{\gamma}\right)^2 + \cdots \right]\cos 2\omega_l t \\
&+ \left[-\frac{1}{24}\left(\frac{V_l}{\gamma}\right)^3 + \cdots \right]\sin 3\omega_l t + \cdots \Bigg\},
\end{aligned} \quad (7.4)
$$

the total instantaneous collector current.

For $(V_l/\gamma) < 1$, (7.4) can be expressed as

$$i_C(t) \simeq I_{CQ}\left(1 + \frac{V_l}{\gamma}\sin \omega_l t\right). \quad (7.5)$$

Raising V_l/γ above unity tends to degrade the noise performance of the mixer [2].

Since C_{be} and C_c are virtually independent of I_C in the micropower range, only $g_{m1} = qI_C/kT$ and $r_{be} = h_{fe0}/g_{m1}$ in the hybrid pi micropower transistor model are modulated substantially by the local oscillator injection voltage v_l. Disregarding for the moment the modulation of r_{be}, the instantaneous value of the output current generator in Figure 7.1b is

$$g_{m1}v_{be}(t) = \frac{i_C(t)}{\gamma}V_s \sin \omega_s t \quad (7.6)$$

for an instantaneous signal voltage

$$v_s(t) = V_s \sin \omega_s t \quad (7.7)$$

appearing across the input terminals of the transistor. Substituting (7.5) into (7.6) yields

$$
\begin{aligned}
g_{m1}v_{be}(t) &= \frac{I_{CQ}}{\gamma}\left[1 + \frac{V_l}{\gamma}\sin \omega_l t\right]V_s \sin \omega_s t \\
&= \frac{I_{CQ}}{\gamma}\left\{V_s \sin \omega_s t + \frac{1}{2}\frac{V_l}{\gamma}V_s[\cos(\omega_l - \omega_s)t - \cos(\omega_l + \omega_s)t]\right\}
\end{aligned} \quad (7.8)
$$

by means of the trigonometric identity $2 \sin A \sin B = \cos(A - B) - \cos(A + B)$. Since the output is tuned to the frequency $\omega_0 = \omega_l - \omega_s$, (7.8) gives

$$g_{m1}v_{be}(t) = \frac{1}{2}\frac{I_{CQ}}{\gamma}\frac{V_l}{\gamma}V_s \cos \omega_0 t \quad (7.9a)$$

for the instantaneous output current generator at ω_0 or

$$g_{m1}V_{be} = \frac{g_m}{2}\frac{qV_l}{kT}V_{be} \qquad (7.9b)$$

for the peak value of the output current generator at ω_0 in terms of the peak value of v_{be} at ω_s. From (7.9b), the conversion transconductance is $g_{m1} \simeq (g_m/2)(qV_l/kT)$, as Figure 7.1b indicates.

The common-emitter diffusion resistance of the transistor operating as a mixer is $r_{be} = h_{fe0}/g_{m1}$ assuming h_{fe0} is substantially independent of I_C. The instantaneous current drawn by this resistor is, using (7.5),

$$i_{be}(t) \simeq \frac{v_{be}}{r_{be}} \simeq \frac{V_s \sin \omega_s t}{h_{fe0}}\frac{I_{CQ}}{\gamma}\left[1 + \frac{V_l}{\gamma}\sin \omega_l t\right]$$

$$\simeq \frac{V_s}{h_{fe0}}\frac{I_{CQ}}{\gamma}\left\{\sin \omega_s t + \frac{V_l}{2\gamma}[\cos(\omega_l - \omega_s)t - \cos(\omega_l + \omega_s)t]\right\} \qquad (7.10)$$

or

$$i_{be}(t) \simeq \frac{V_s}{h_{fe0}}\frac{I_{CQ}}{\gamma}\frac{V_l}{2\gamma}\cos \omega_0 t \qquad (7.11)$$

at the output frequency ω_0. Assuming the internal impedance of the signal source and local oscillator are essentially zero at ω_0, the full voltage drop due to $i_{be}(t)$ of (7.11) must appear across the transistor base resistance r_b (heretofore neglected). Consequently, at ω_0 for $r_b \ll |1/j\omega_0(C_{be} + C_c)|$

$$V_{be}(t) = -i_{be}r_b \qquad (7.12a)$$

and

$$V_{be} = -\left(\frac{V_s}{h_{fe0}}\frac{qI_{CQ}}{kT}\right)\left(\frac{1}{2}\frac{qV_l}{kT}\right)r_b \qquad (7.12b)$$

gives the peak value.

For $qV_l/kT \simeq 1$, the ratio of peak voltages

$$\left|\frac{V_{be}}{V_s}\right| \simeq \left(\frac{1}{h_{fe0}}\frac{qI_{CQ}}{kT}\right)\frac{r_b}{2} \ll 1. \qquad (7.13)$$

Therefore the component of the output current at ω_0 due to amplification of $V_{be}(\omega_0)$ of (7.12) is much smaller than the component of the output current due to mixing as described by (7.9). The simple model of Figure 7.1b is sufficient then to describe accurately mixer operation in the micropower range for $qV_l/kT < 1$.

Mixer Noise Figure

Superimposing the uncorrelated noise sources described by (6.2) and (6.3) on the model of Figure 7.1b, the noise figure of the mixer [4,5,6] can be calculated. The effect at the output of the mixer of all of the noise sources in the transistor for a narrow frequency interval around ω_s can be represented by a single equivalent noise generator $\overline{e_{eq}^2}$ located in series with the source. The mean square value of this equivalent noise generator is

$$\overline{e_{eq}^2} = \overline{e_b^2} + \overline{i_b^2}(R_s + r_b)^2 + \overline{i_c^2}\frac{|R_s + r_b + Z_{be}|^2}{|g_{m1}Z_{be}|^2}, \qquad (7.14)$$

assuming

$$|Z_{be}| = \left|\frac{r_{be}}{1 + j\omega_s r_{be} C_{be}}\right| \ll \left|\frac{1}{j\omega_s C_c}\right|,$$

as derived from Figure 6.1 and Figure 7.1b combined. The noise figure F of the mixer is defined as the ratio of the total noise power delivered to R_o (at ω_0) to the noise power that would be delivered to R_o (at ω_0) if the only noisy component were the source resistance R_s at a temperature $T_0 = 290°\text{K}$. An approximate expression for F is

$$F = \frac{\overline{e_{eq}^2} + \overline{e_s^2}}{\overline{e_s^2}}, \qquad (7.15)$$

where $\overline{e_s^2}$ represents the mean square source noise in a narrow frequency interval about ω_s. For $\overline{e_s^2} = 4kT_0 R_s \Delta f$, substitution of (6.2), (6.3), and (7.14) into (7.15) gives

$$F = 1 + \frac{r_b}{R_s} + \frac{1}{2\gamma}\frac{I_C}{h_{FE}}\frac{(R_s + r_b)^2}{R_s}$$
$$+ \frac{1}{2\gamma}\frac{I_C}{h_{fe0}^2}\frac{(R_s + r_b + r_{be})^2}{R_s}\left(\frac{g_m}{g_{m1}}\right)^2 + \frac{1}{2\gamma}I_C\frac{(R_s + r_b)^2}{R_s}\left(\frac{f_s}{f_T}\right)^2\left(\frac{g_m}{g_{m1}}\right)^2, \qquad (7.16)$$

where

$$\frac{g_m}{g_{m1}} \simeq 2\frac{kT}{qV_l} \qquad \text{for} \qquad \frac{qV_l}{kT} < 1,$$

as described by (7.9). This result compares closely with the amplifier noise figure (6.7), but neglects the output noise from all possible conversion frequencies except the signal frequency band and thus represents an optimistic value for F. However, the result is useful in determining the minimum quiescent power for the mixer, particularly since the signal frequency band noise contribution is larger than that of any other band.

Assuming the frequency dependent term in (7.16) is negligible because of proper selection of the dc operating point of the transistor, $\partial F/\partial R_s$ indicates that a minimum value of the frequency independent portion of (7.16)

$$F_0(\min)\bigg|_{I_c=\text{constant}} = 1 + \left[\frac{(g_m/g_{m1})^2 + 2g_m r_b}{h_{FE}}\right]^{1/2} \quad (7.17)$$

occurs for an optimum value of source resistance given by

$$R_s^2 \simeq h_{FE} \frac{(g_m/g_{m1})^2 + 2g_m r_b}{g_m^2}. \quad (7.18)$$

Substituting (7.18) into (7.16) yields

$$F \simeq 1 + \left[\frac{(g_m/g_{m1})^2 + 2g_m r_b}{h_{FE}}\right]^{1/2}\left[1 + \frac{h_{FE}}{2}\left(\frac{\omega_s}{\omega_T}\right)^2\left(\frac{g_m}{g_{m1}}\right)^2\right]. \quad (7.19)$$

Comparing (6.24) and (7.19) the noise figure degradation that results in a mixer due to $(g_m/g_{m1}) = 2(kT/qV_t) > 2$ is evident. If the frequency dependent term in (7.19) is restricted to increasing F-1 by 3 dB, the relationship between the signal frequency ω_s and the transistor gain-bandwidth product ω_T must satisfy the equation

$$\omega_s^2 = \frac{2}{h_{FE}}\left(\frac{g_{m1}}{g_m}\right)^2\left(\frac{g_m}{C_{be} + C_c}\right)^2. \quad (7.20)$$

This result in turn determines the minimum quiescent power of the mixer

$$P_{DC}^* = V_{CC}I_C \simeq \gamma V_{CC}\left(\frac{h_{FE}}{2}\right)^{1/2}(C_{be} + C_c)\left(\frac{g_m}{g_{m1}}\right)\omega_s. \quad (7.21)$$

It has frequently been observed that a transistor mixer can be analyzed as if it were a (base-emitter) diode in which mixing occurs followed by a transistor amplifier at the intermediate frequency ω_0 [3,5,6,7]. However, from (7.21) it is evident that for minimum noise figure the quiescent power required by the mixer is proportional to its input signal frequency, whereas the quiescent power required by a corresponding low-noise intermediate frequency amplifier is proportional to the output or intermediate frequency as indicated by (6.26).

Conversion Gain

Based on Figure 7.1, a conversion voltage gain [3] for the mixer may be expressed as $A_{vc} = v_0(\omega_0)/e_s(\omega_s)$, where v_0 is the voltage across R_o. Thus

$$A_{vc} = -\frac{g_m}{2}\left(\frac{qV_l}{kT}\right)R_o\left(\frac{r_{be}}{R_s + r_{be}}\right) \quad (7.22a)$$

$$\simeq -\frac{g_m}{2}\left(\frac{qV_l}{kT}\right)R_o, \quad (7.22b)$$

if R_s is described by (7.18). Because the output of the mixer is tuned to ω_0 while the input tank circuits are tuned to the local oscillator frequency ω_l and the signal frequency ω_s, instability due to internal transistor feedback is not as serious a problem as in a tuned amplifier.

7.2 A MONOLITHIC MIXER

Figure 5.7 illustrates the schematic diagram of two micropower mixers utilizing monolithic integrated circuits (enclosed within the dashed rectangles). Figure 7.2 shows a photomicrograph of the monolithic chip of the mixers. The salient performance characteristics of the mixers are listed [8]:

	First Mixer	Second Mixer
power gain	18–20 dB	25–27 dB
noise figure	9–11 dB	15–17 dB
signal frequency	51 MHz	10.7 MHz
local oscillator frequency	40.3 MHz	10.245 MHz
intermediate frequency	10.7 MHz	455 kHz
nominal supply voltage	3.0 V	3.0 V
power drain	900 μW	900 μW

A transistor with $f_T \simeq 450$ MHz at $I_C = 5.0$ ma, $V_{CE} = 5.0$ V was used at a reduced current level to obtain the low-power drain of 900 μW. For $I_C = 300$ μA the upper noise corner frequency for the transistor is approximately 25 MHz.

7.3 DETECTORS [1,2]

The circuit diagrams of a transistor amplitude modulation detector or demodulator are illustrated in Figure 7.3 [6]. The design constraints for such a detector may include conversion gain, signal frequency ω_s, bandwidth Δf, modulation frequency ω_m, as well as other requirements. A quasilinear analysis to determine the minimum quiescent power of a detector is discussed in this section.

From (7.1) and (7.3) the instantaneous transistor collector current can be written as

$$i_C(t) = I_{CQ}\left[1 + \frac{e_s(t)}{\gamma} + \frac{1}{2!}\left(\frac{e_s(t)}{\gamma}\right)^2 + \frac{1}{3!}\left(\frac{e_s(t)}{\gamma}\right)^3 + \cdots\right], \quad (7.23)$$

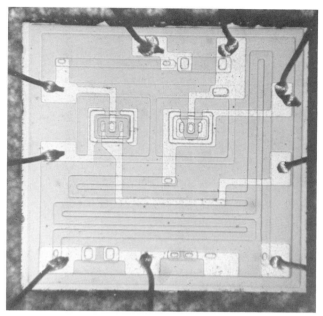

Figure 7.2 Photomicrograph of the monolithic integrated circuit used in the mixers of Figure 5.7. Resistor line width is 1 mil, chip is 40 × 40 mils (courtesy of U.S. Army Electronics Command).

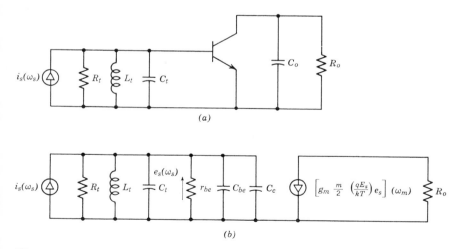

Figure 7.3 Circuit diagrams for transistor AM detector: (a) ac schematic diagram; (b) equivalent circuit.

117

where
$$e_s(t) = E_s(1 + m \sin \omega_m t) \sin \omega_s t \qquad (7.24)$$

represents the amplitude modulated signal voltage across the input terminals of the transistor.

Substituting (7.24) into (7.23) gives

$$\begin{aligned}
i_C(t) = I_{CQ} \Bigg\{ & \left[1 + \frac{1}{4}\left(\frac{E_s}{\gamma}\right)^2 (1 + m \sin \omega_m t)^2 \right. \\
& \left. + \frac{1}{4} \cdot \frac{1}{16}\left(\frac{E_s}{\gamma}\right)^4 (1 + m \sin \omega_m t)^4 + \cdots \right] \\
& + \left[\frac{E_s}{\gamma}(1 + m \sin \omega_m t) + \frac{1}{2 \cdot 4}\left(\frac{E_s}{\gamma}\right)^3 (1 + m \sin \omega_m t)^3 + \cdots \right] \sin \omega_s t \\
& + \left[-\frac{1}{4}\left(\frac{E_s}{\gamma}\right)^2 (1 + m \sin \omega_m t)^2 + \cdots \right] \cos 2\omega_s t + \cdots \Bigg\}. \qquad (7.25)
\end{aligned}$$

Assuming $\omega_s \gg \omega_m$, the output capacitor C_o provides a very low impedance path for currents at ω_s, its harmonics and mixing products. Consequently, neglecting all but the first term, (7.25) can be written as

$$i_C(t) = I_{CQ} \left[1 + \frac{1}{4}\left(\frac{E_s}{\gamma}\right)^2 (1 + m \sin \omega_m t)^2 \right. \\
\left. + \frac{1}{4} \cdot \frac{1}{16}\left(\frac{E_s}{\gamma}\right)^4 (1 + m \sin \omega_m t)^4 + \cdots \right]. \qquad (7.26)$$

In the "small" signal case where $E_s/\gamma < 1$, (7.26) gives

$$i_C(t) \simeq I_{CQ} \left[1 + \frac{1}{4}\left(\frac{E_s}{\gamma}\right)^2 (1 + 2m \sin \omega_m t) \right] \qquad (7.27)$$

for $m \ll 1$. From (7.27),

$$I_c(\omega_m) = I_{CQ} \frac{m}{2}\left(\frac{E_s}{\gamma}\right)^2 \qquad (7.28)$$

gives the peak value of I_c at ω_m.

In the micropower range C_{be} is virtually independent of e_s so that C_{be}, C_c, and C_t resonate with L_t at ω_s. Since the second term of (7.25) yields $I_c(\omega_s) = I_{CQ}(E_s/\gamma)$, the peak value of I_c at ω_s,

$$I_s(\omega_s) = \left[\frac{1/r_{be} + 1/R_T}{1/r_{be}}\right] \frac{1}{h_{fe0}} I_{CQ} \left(\frac{E_s}{\gamma}\right) \qquad (7.29)$$

is the peak value of the signal current source. From (7.28) and (7.29) the conversion current gain is

$$A_{ic} = \frac{I_c(\omega_m)}{I_s(\omega_s)} \tag{7.30a}$$

$$= \frac{h_{fe0}}{2} m \left(\frac{qE_s}{kT}\right) \frac{1/r_{be}}{1/r_{be} + 1/R_t}. \tag{7.30b}$$

The bandwidth of the input circuit is

$$\Delta f \simeq \frac{1/r_{be} + 1/R_t}{C_{be} + C_c + C_t} = \frac{\omega_s}{Q} \tag{7.31}$$

and the quiescent power of the detector can be expressed as

$$P_{DC}^* = V_{CC} I_C \tag{7.32a}$$

$$= 2\gamma^2 V_{CC} \frac{A_{ic}}{mE_s} (C_{be} + C_c + C_t) \Delta f \tag{7.32b}$$

by substitution of (7.30b) and (7.31) into (7.32a). This result describes the power drain of the detector for (a) a small signal input, $qE_s/kT < 1$, (b) a small index of modulation, $m \ll 1$, and (c) the input susceptance of the transistor tuned out by the tank circuit.

For a micropower detector, only the third of these three conditions is satisfied approximately in most instances. For $(E_s/\gamma) > 1$ and $m \to 1$, the first and second terms in (7.25) give

$$i_C(t) = I_{CQ} \left\{ \left[1 + \frac{1}{4}\left(\frac{E'_s}{\gamma}\right)^2 + \frac{1}{4} \cdot \frac{1}{16}\left(\frac{E'_s}{\gamma}\right)^4 + \cdots \right] \right.$$

$$\left. + \left[\frac{E'_s}{\gamma} + \frac{1}{2} \cdot \frac{1}{4}\left(\frac{E'_s}{\gamma}\right)^3 + \cdots \right] \sin \omega_s t \right\}, \tag{7.33}$$

where

$$E'_s = E_s(1 + m \sin \omega_m t) \tag{7.34a}$$

and

$$i'_C \equiv I_{CQ} \left[1 + \frac{1}{4}\left(\frac{E'_s}{\gamma}\right)^2 + \frac{1}{4} \cdot \frac{1}{16}\left(\frac{E'_s}{\gamma}\right)^4 + \cdots \right]. \tag{7.34b}$$

From (7.33),

$$I_s(\omega_s) = \frac{1/r_{be} + 1/R_T}{1/r_{be}} \frac{1}{h_{fe0}} I_{CQ} \left[\frac{E'_s}{\gamma} + \frac{1}{2} \cdot \frac{1}{4}\left(\frac{E'_s}{\gamma}\right)^3 + \cdots \right] \tag{7.35}$$

so that under large signal conditions, $i'_C \gg I_{CQ}$,

$$A_{ic} = \frac{i'_C - I_{CQ}}{I_s(\omega_s)} \tag{7.36a}$$

$$= \frac{h_{fe0}}{4} \frac{qE'_s}{kT} \frac{1/r_{be}}{1/r_{be} + 1/R_t} \left\{ \frac{\frac{E'_s}{\gamma} + \frac{1}{16}\left(\frac{E'_s}{\gamma}\right)^3 + \frac{1}{16 \cdot 36}\left(\frac{E'_s}{\gamma}\right)^5 + \cdots}{\frac{E'_s}{\gamma} + \frac{1}{8}\left(\frac{E'_s}{\gamma}\right)^3 + \frac{1}{12 \cdot 16}\left(\frac{E'_s}{\gamma}\right)^5 + \cdots} \right\} \tag{7.36b}$$

$$\simeq \frac{h_{fe0}}{2} \frac{qE'_s}{kT} \frac{1/r_{be}}{1/r_{be} + 1/R_t}, \tag{7.36c}$$

which can be compared with (7.30b).

REFERENCES

[1] R. F. Shea et al., *Transistor Circuit Engineering*, Wiley, New York, 1957, Chapter 9.
[2] D. Dewitt and A. L. Rossoff, *Transistor Electronics*, McGraw-Hill, New York, 1957, Chapter 12.
[3] J. Zawels, "The Transistor as a Mixer." *Proc. IRE*, **42**, 542–548 (March 1954).
[4] J. S. Vogel and M. J. O. Strutt, "Das Rauschen in Transistormischstufen," *Archiv Elekt. Ubertragung*, **16**, 215–222 (May 1962).
[5] R. R. Webster, "The Noise Figure of Transistor Converters," *IRE Trans. Broadcast Television Receivers*, **7**, 50–65 (November 1961).
[6] J. D. Meindl and P. H. Hudson, "Low Power Linear Circuits," *IEEE J. Solid-State Circuits*, **1**, 100–111 (December 1966).
[7] Engineering Staff of Texas Instruments, Inc., "Transistor Circuit Design," McGraw-Hill, New York, 1963, Chapter 23.
[8] J. Feit and R. McGinnis, "Small Signal Function Circuit Units," USAECOM, Fort Monmouth, N.J., Contract No. DA 28-043 AMC-00151 (E), Final Report, July 1964–June 1965.

Chapter 8

Feedback Amplifiers

The principal advantages of negative feedback in an amplifier [1–4] are (a) a reduction in the sensitivity of the gain of the amplifier to parameter variations of its elements, (b) a reduction of output signal distortion due to nonlinearity in the amplifier characteristics, and (c) an increase in the bandwidth of the amplifier. Amplifier gain is exchangeable for these advantages provided that self-oscillation or instability is avoided in the design. The advantages of negative feedback in micropower amplifiers are discussed in this chapter. Since the principal advantage is improved desensitivity, this topic is emphasized and distortion and bandwidth are treated more briefly to add perspective.

8.1 SENSITIVITY

The block diagram of an elementary feedback amplifier is illustrated in Figure 8.1. Assuming that the basic amplifier a is unilateral and that forward transmission through a is much larger than through the feedback network f, the closed-loop gain is given by

$$A_v = \frac{V_0}{V_i} = \frac{a}{1 + af}. \tag{8.1}$$

For an iterative cascade of n stages of these elementary feedback amplifiers the overall amplifier gain is

$$A_v^* = A_v^{\,n} = \left(\frac{a}{1 + af}\right)^n. \tag{8.2}$$

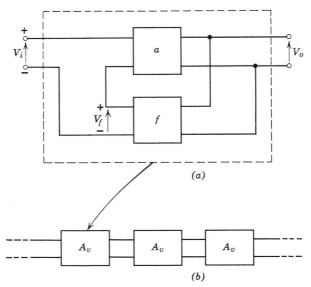

Figure 8.1 (a) Elementary feedback amplifier block diagram; (b) iterative chain of elementary feedback amplifiers.

Taking the differential of (8.2) yields

$$\frac{dA_v^*}{A_v^*} = n\frac{dA_v}{A_v} = n\frac{1}{1+af}\left(\frac{da}{a} - a\,df\right), \tag{8.3}$$

the relative change in overall gain for a small fractional change in the basic amplifier and the feedback network transfer functions.

For large changes Δa in a but negligible changes in f, (8.2) gives

$$\frac{\Delta A_v}{A_v} = \frac{1}{1+(a+\Delta a)f}\frac{\Delta a}{a} \tag{8.4}$$

and

$$\frac{\Delta A_v^*}{A_v^*} = \left(1 + \frac{\Delta A_v}{A_v}\right)^n - 1$$

$$= n\frac{\Delta A_v}{A_v} + \frac{n(n-1)}{2!}\left(\frac{\Delta A_v}{A_v}\right)^2 + \frac{n(n-1)(n-2)}{3!}\left(\frac{\Delta A_v}{A_v}\right)^3 + \cdots. \tag{8.5}$$

In both (8.3) and (8.4) it is evident that the fractional change in a (i.e., da/a or $\Delta a/a$) is diminished by the factor $(1+af)^{-1}$ or $[1+(a+\Delta a)f]^{-1}$ in its effect on the fractional change in A_v ($af = T$ is the loop gain of the feedback amplifier).

Sensitivity

If the feedback loop illustrated in Figure 8.1a includes n basic amplifier stages each with a unilateral gain a, the overall closed-loop gain is given by

$$A'_v = \frac{V_0}{V_i} = \frac{a^n}{1 + a^n F} \tag{8.6}$$

and

$$\frac{dA'_v}{A'_v} = \frac{n}{1 + a^n F} \frac{da}{a} - A'_v \, dF. \tag{8.7}$$

Consider (8.3) and (8.7) for $df = 0$ and $dF = 0$,

$$\frac{dA'_v/A'_v}{dA^*_v/A^*_v} = \frac{1 + af}{1 + a^n F}. \tag{8.8}$$

Since $A^*_v = A'_v$, $A^*_v \simeq (1/f)^n$ and $A'_v \simeq 1/F$, (8.8) gives

$$\frac{\delta A'_v}{\delta A^*_v} \simeq \left(\frac{1}{af}\right)^{n-1} \ll 1, \tag{8.9}$$

where the differentials dA'_v and dA^*_v have been replaced by the small finite changes $\delta A'_v$ and δA^*_v, respectively. From (8.9) it is evident that the multistage feedback loop can be more effective than single stage feedback in reducing sensitivity.

Local Emitter Feedback

For the iterative amplifier chain with local emitter feedback [5] illustrated in Figure 8.2, the voltage gain per stage in the midband range is

$$A_v = -\frac{V_{1(n+1)}}{V_{1n}} = \frac{h_{fe0} R_c/(r_{be} + R_e)}{1 + h_{fe0} R_c/(r_{be} + R_e)(R_e/R_c)} \tag{8.10}$$

when written in the form $A_v = a/(1 + af)$ where $a = h_{fe0} R_c/(r_{be} + R_e)$ and $f = R_e/R_c$. The differential of the basic amplifier gain per stage is

$$da = \frac{\partial a}{\partial h_{fe0}} dh_{fe0} + \frac{\partial a}{\partial g_m} dg_m + \frac{\partial a}{\partial R_c} dR_c \tag{8.11}$$

so that (8.10) gives

$$\frac{da}{a} = \frac{a}{h_{fe0}} \frac{dh_{fe0}}{h_{fe0}} + \frac{a}{g_m R_c} \frac{dg_m}{g_m} + \frac{a}{g_m R_c} \frac{dR_c}{R_c}. \tag{8.12}$$

For the feedback factor,

$$df = \frac{\partial f}{\partial R_e} dR_e + \frac{\partial f}{\partial R_c} dR_c \tag{8.13}$$

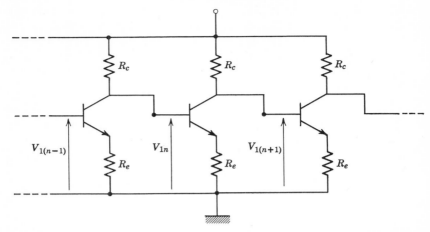

Figure 8.2 AC schematic diagram of iterative amplifier chain with local emitter feedback.

and from (8.10)

$$\frac{df}{f} = \frac{dR_e}{R_e} - \frac{dR_c}{R_c}. \tag{8.14}$$

Combining (8.3), (8.12), and (8.14) gives

$$\frac{dA_v^*}{A_v^*} = \frac{n}{1+af}\left[\left(\frac{a}{h_{feo}}\frac{dh_{feo}}{h_{feo}} + \frac{a}{g_mR_c}\frac{dg_m}{g_m} + \frac{a}{g_mR_c}\frac{dR_c}{R_c}\right) - af\left(\frac{dR_e}{R_e} - \frac{dR_c}{R_c}\right)\right]. \tag{8.15}$$

In many instances the tolerance of h_{feo} will dominate (8.15). In such cases a useful approximation for the lower bound of the fractional change in overall gain is

$$\frac{dA_v^*}{A_v^*} = n\frac{(A_v^*)^{1/n}}{h_{feo}}\frac{dh_{feo}}{h_{feo}} \tag{8.16a}$$

from (8.15). The sensitivity of A_v^* to changes in h_{feo} is

$$S_E = \frac{dA_v^*/A_v^*}{dh_{feo}/h_{feo}}. \tag{8.16b}$$

From (8.16b), $\partial S_E/\partial n = 0$ yields

$$n\,(\text{optimum}) = \ln A_v^* \tag{8.17a}$$

the optimum number of stages and

$$A_v\,(\text{optimum}) = \epsilon = 2.7, \tag{8.17b}$$

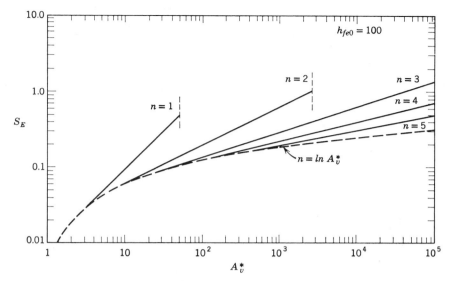

Figure 8.3 Sensitivity, $S_E = (dA_v^*/A_v^*)/(dh_{fe0}/h_{fe0})$ versus overall amplification for an iterative amplifier chain with local emitter feedback.

the optimum gain per stage for minimum sensitivity. Substituting (8.17) into (8.16a) gives

$$\frac{dA_v^*}{A_v^*} \text{(minimum)} = \ln A_v^* \frac{\epsilon}{h_{fe0}} \frac{dh_{fe0}}{h_{fe0}}. \quad (8.18)$$

This result indicates the following:

1. There is a minimum possible value for the sensitivity of A_v^* to changes in h_{fe0}.
2. This minimum value is independent of R_e, although the effect on dA_v^*/A_v^* of tolerances in g_m and R_c is reduced as R_e increases, since (8.15) indicates

$$\frac{1}{1+af} \frac{a}{g_m R_c} \left(\frac{dg_m}{g_m} + \frac{dR_c}{R_c} \right) \simeq \frac{1}{g_m R_e} \left(\frac{dg_m}{g_m} + \frac{dR_c}{R_c} \right).$$

3. As h_{fe0} increases, S_E is reduced (assuming dh_{fe0}/h_{fe0} remains fixed) as a consequence of greater interstage impedance mismatch (that is, the increase in $r_{be} + h_{fe0}R_e$ relative to R_c).
4. The small gain per stage required for minimum sensitivity is compatible with the design approach for minimum quiescent power as indicated, for example, by (4.23).

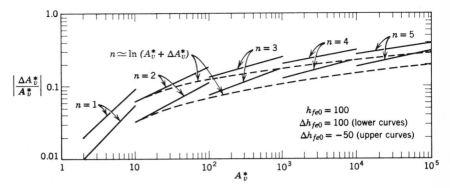

Figure 8.4 Fractional change in overall gain $\Delta A_v^*/A_v^*$ versus overall gain A_v^* for an iterative amplifier chain with local emitter feedback.

A graphical representation of (8.16a) and (8.18) with $h_{fe0} = 100$ is illustrated in Figure 8.3. Although the curves are based upon the assumption that the changes in h_{fe0} are small, they are an approximate indication that the fractional change in the total gain of a multistage amplifier $\Delta A_v^*/A_v^*$ can be limited to less than about 30% for $|\Delta h_{fe0}/h_{fe0}| \leq 1$ for each stage of the amplifier.

For large changes in h_{fe0} (8.5) in conjunction with

$$\frac{\Delta A_v}{A_v} = \frac{A_v + \Delta A_v}{h_{fe0} + \Delta h_{fe0}} \frac{\Delta h_{fe0}}{h_{fe0}} \tag{8.19}$$

gives

$$\frac{\Delta A_v^*}{A_v^*} = n \frac{A_v + \Delta A_v}{h_{fe0} + \Delta h_{fe0}} \left[1 + \frac{n-1}{2!} \frac{\Delta A_v}{A_v} + \frac{(n-1)(n-2)}{3!} \left(\frac{\Delta A_v}{A_v}\right)^2 + \cdots \right] \frac{\Delta h_{fe0}}{h_{fe0}}. \tag{8.20}$$

Figure 8.4 illustrates a graphical representation of (8.20). Assuming the worst possible tolerance for h_{fe0} in each stage of the amplifier, Figure 8.4 predicts that the fractional change in overall amplification lies within about $\pm 30\%$ of the nominal value.

Local Emitter and Collector Feedback

The ac schematic diagram of an iterative amplifier chain with both emitter and collector feedback [5–7] is illustrated in Figure 8.5. The voltage gain per

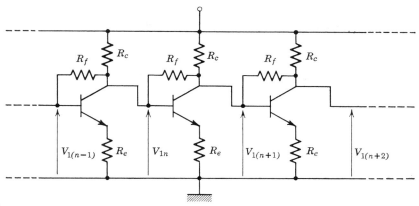

Figure 8.5 AC schematic diagram of iterative amplifier chain with local emitter and collector feedback.

stage in the midband range is

$$A_{vn} = -\frac{V_{1(n+1)}}{V_{1n}} = \frac{\dfrac{h_{fe0}R_c}{r_{be}(1 + g_m R_e) + R_c}}{1 + \dfrac{h_{fe0}R_c}{r_{be}(1 + g_m R_e) + R_c}\dfrac{(1 + g_m R_e)}{g_m R_f}[1 + A_{v(n+1)}]}.$$

(8.21)

In those instances in which the tolerance of h_{fe0} dominates the sensitivity of (8.21) the lower bound on the fractional change in A_{vn} is

$$\frac{dA_{vn}}{A_{vn}} \simeq A_{vn}\left[\frac{1}{h_{fe0}}\frac{dh_{fe0}}{h_{fe0}} - f_n\frac{dA_{v(n+1)}}{A_{v(n+1)}}\right], \quad (8.22)$$

where

$$f_n = \frac{(1 + g_m R_e)}{g_m R_f}[1 + A_{v(n+1)}] \quad \text{and} \quad A_{v(n+1)} = -\frac{V_{1(n+2)}}{V_{1(n+1)}}.$$

From (8.22)

$$\frac{dA_{vn}}{A_{vn}} + \frac{dA_{v(n+1)}}{A_{v(n+1)}} \simeq \frac{A_{vn}}{h_{fe0}}\frac{dh_{fe0}}{h_{fe0}} \quad (8.23)$$

for $af_n \gg 1$ with $a = h_{fe0}R_c/[r_{be}(1 + g_m R_e) + R_c]$. Combining (8.3) and (8.23) gives

$$\frac{dA_v^*}{A_v^*} = \frac{n}{2}\frac{(A_v^*)^{1/n}}{h_{fe0}}\frac{dh_{fe0}}{h_{fe0}}, \quad (8.24)$$

which is comparable to (8.16). The optimum number of stages,

$$n \text{ (optimum)} = 2 \ln A_{v0}^*, \quad (8.25)$$

the optimum gain per stage,

$$A_v \text{ (optimum)} = \sqrt{\epsilon} = 1.65, \tag{8.26}$$

and the minimum value of dA_v^*/A_v^*,

$$\frac{dA_v^*}{A_v^*} \text{ (minimum)} = \ln A_v^* \frac{\sqrt{\epsilon}}{h_{fe0}} \frac{dh_{fe0}}{h_{fe0}}, \tag{8.27}$$

can be derived by differentiation of (8.24). Comparing (8.18) and (8.27) shows an improved sensitivity for the circuit of Fig. 8.5.

Alternate Emitter and Collector Feedback Stages [5]

The voltage gain of the iterative pair of stages enclosed within the dashed block of Figure 8.6 is

$$A_{vp} = \frac{a_{v1} a_{v2}}{1 + a_{v1} a_{v2} f_p}, \tag{8.28}$$

where

$$a_{v1} = \frac{h_{fe0} R_{c1}}{r_{be2} + R_{c1}} \tag{8.29a}$$

$$a_{v2} = \frac{h_{fe0} R_{c2}}{r_{be1}(1 + g_{m1} R_e) + R_{c2}} \tag{8.29b}$$

and

$$f_p = \frac{1 + g_{m1} R_e}{g_{m1} R_f} \tag{8.29c}$$

for $a_{v2} \gg 1$. The fractional change in the overall gain of n_p of these pairs is

$$\frac{dA_v^*}{A_v^*} = n_p \frac{dA_{vp}}{A_{vp}} = \frac{n_p}{1 + a_{v1} a_{v2} f_p} \left(\frac{da_{v1}}{a_{v1}} + \frac{da_{v2}}{a_{v2}} \right) \tag{8.30}$$

for $A_v^* = A_{vp}{}^{n_p}$ and $df_p \simeq 0$. Since

$$\frac{da_{v1}}{a_{v1}} \simeq \frac{a_{v1}}{h_{fe0}} \frac{dh_{fe0}}{h_{fe0}} \quad \text{and} \quad \frac{da_{v2}}{a_{v2}} \simeq \frac{a_{v2}}{h_{fe0}} \frac{dh_{fe0}}{h_{fe0}} \tag{8.31}$$

when the fractional changes in a_{v1} and a_{v2} are dominated by dh_{fe0}/h_{fe0}, (8.30) gives

$$\frac{dA_v^*}{A_v^*} = n_p \left(\frac{1}{a_{v1}} + \frac{1}{a_{v2}} \right) \frac{(A_v^*)^{1/n_p}}{h_{fe0}} \frac{dh_{fe0}}{h_{fe0}}. \tag{8.32}$$

A comparison of the sensitivities of the iterative emitter feedback cascade (8.16a) and the alternate stage cascade (8.32) for an equal number of stages

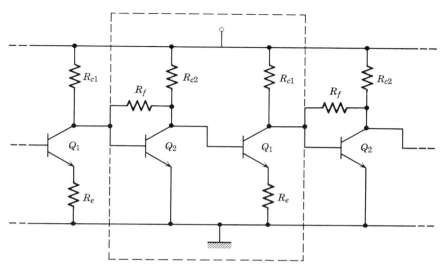

Figure 8.6 AC schematic of alternate emitter and collector feedback stages.

(i.e., $n = 2n_p$) and the same overall gain A_v^* gives

$$\frac{S(\text{alternate stage cascade})}{S(\text{emitter feedback cascade})} \equiv \frac{S_A}{S_E} \simeq \frac{\sqrt{A_{vp}}}{2a_{v2}} \qquad (8.33a)$$

for $a_{v1} \gg a_{v2}$. For $A_{vp} \simeq R_f/R_e > a_{v2} \simeq R_{c2}/R_e$ (8.33a) gives

$$\frac{S_A}{S_E} \simeq \frac{1}{2}\left[\frac{R_f/R_e}{(R_{c2}/R_e)^2}\right]^{1/2} < 1 \qquad (8.33b)$$

for $A_{vp} \lesssim 4a_{v2}^2$.

On the basis of (8.33), it is evident that the sensitivity of the overall gain of the alternate stage cascade in Figure 8.6 to changes in h_{fe0} can be several times better than the corresponding sensitivity of the emitter feedback cascade in Figure 8.2. This improvement in sensitivity results from (a) the increased interstage impedance mismatch of the alternate stage cascade compared to the emitter feedback cascade and (b) the fact that the voltage gain of the alternate pair is the product of the two most stable transfer functions of its two stages—the transadmittance transfer function of the emitter feedback stage and the transresistance transfer function of the collector feedback stage.

Multistage Feedback

On the basis of Figure 8.3, Figure 8.4, and (8.33) we can expect to constrain the gain of an alternate stage cascade to within approximately ±10% of the design value for a 4:1 tolerance, $0.5h_{feo} \leq h_{feo} \leq 2h_{feo}$, on transistor current gain. This amount of desensitivity will suffice for the majority of micropower applications. Where still smaller fractional changes in A_v^* are required, multistage feedback loops can be employed in conjunction with the well-stabilized local feedback stages discussed in this section.

8.2 DISTORTION [1–4,8]

Figure 8.7 illustrates the block diagram of an elementary feedback amplifier in which an extraneous signal V_x arises because of a nonlinearity within the basic amplifier. The output voltage is

$$V_0 = \frac{a_1 a_2}{1 + a_1 a_2 f}\left(V_1 + \frac{V_x}{a_1}\right). \tag{8.34a}$$

For an output distortion voltage V_{0x} and an output signal voltage V_{0s}, where $V_0 = V_{0x} + V_{0s}$, (8.34a) gives

$$\frac{V_{0x}}{V_{0s}} = \frac{a_2}{1 + a_1 a_2 f}\frac{V_x}{V_{0s}}$$

$$\simeq \frac{1}{a_1 f}\frac{V_x}{V_{0s}}, \quad a_1 a_2 f \gg 1 \tag{8.34b}$$

$$\simeq a_2 \frac{V_x}{V_{0s}}, \quad f = 0$$

the output distortion-to-signal ratio. Clearly, for constant output signal voltage V_{0s},

$$\frac{V_{ox}/V_{os}|_{a_1 a_2 f \gg 1}}{V_{ox}/V_{os}|_{f=0}} = \frac{1}{a_1 a_2 f} \ll 1 \tag{8.34c}$$

which indicates the improvement in output distortion-to-signal ratio resulting from negative feedback.

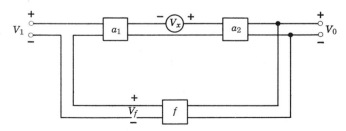

Figure 8.7 Elementary feedback amplifier block diagram illustrating a distortion voltage V_x.

8.3 BANDWIDTH [1–4]

If the frequency dependence of the gain of the basic amplifier of Figure 8.1a is described by

$$a = \frac{a_0}{1 + j(\omega/\omega_{ca})}, \quad (8.35)$$

the closed-loop gain is given by

$$A_v = \frac{A_{v0}}{1 + j\omega/(1 + a_0 f)\omega_{ca}}, \quad (8.36)$$

where $A_{v0} = a_0/(1 + a_0 f)$ and f is assumed independent of frequency. From (8.36), the 3 dB bandwidth of the closed loop amplifier stage is

$$\omega_c = (1 + a_0 f)\omega_{ca}, \quad (8.37a)$$

and its gain-bandwidth product is

$$A_{v0}\omega_c = a_0\omega_{ca}, \quad (8.37b)$$

which indicates an ideal one-to-one trade-off of gain for bandwidth as feedback is applied to the basic amplifier. For an iterative cascade of n stages,

$$A_v^* = A_v^n = \left(\frac{A_{v0}}{1 + j(\omega/\omega_c)}\right)^n \quad (8.38)$$

so that

$$A_{v0}^* = A_{v0}^n \quad \text{and} \quad \omega_c^* = (2^{1/n} - 1)^{1/2}\omega_c. \quad (8.39)$$

For the emitter and collector feedback iterative amplifier chain in Figure 8.5,

$$A_v \simeq \frac{A_{v0}}{1 + j\omega A_{v0} \frac{(1 + g_m R_e)}{g_m}\left[\frac{C_{be}}{1 + g_m R_e} + (1 + A_{v0})C_c\right]} \quad (8.40a)$$

where

$$A_{v0} = \frac{\frac{h_{feo}R_c}{r_{be}(1 + g_m R_e) + R_c}}{1 + \frac{h_{feo}R_c}{r_{be}(1 + g_m R_e) + R_c}\frac{(1 + g_m R_e)}{g_m R_f}(1 + A_{v0})} \quad (8.40b)$$

$$\omega_c = \frac{1}{A_{v0}}\frac{g_m}{C_{be} + (1 + g_m R_e)(1 + A_{v0})C_c} \quad (8.40c)$$

and

$$A_{v0}\omega_c = \frac{g_m}{C_{be} + (1 + g_m R_e)(1 + A_{v0})C_c}. \quad (8.40d)$$

When we compare (8.40) with (4.8), (4.9), and (4.10), it is evident that the ideal behavior of a feedback amplifier postulated in the preceding paragraph is approximately followed by the circuit in Figure 8.5. Some loss of gain-bandwidth product does accompany the introduction of emitter feedback. Collector feedback has the same influence on gain-bandwidth product as resistive broadbanding of micropower amplifiers.

8.4 STABILITY [1–4]

In the elementary feedback amplifier in Figure 8.1, both a and f are assumed to be real negative numbers in the midband frequency range. In terms of the complex frequency, $s = \sigma + j\omega$, the closed-loop gain of the elementary feedback amplifier is

$$A_v(s) = \frac{a(s)}{1 + a(s)f(s)}. \tag{8.41}$$

The requirement for stable amplification is that none of the poles of $A_v(s)$ lie on the $j\omega$ axis or in the right half of the complex frequency plane. If, for example,

$$a(s) = \frac{a(0)}{1 + a_1 s + a_2 s^2 + \cdots + a_n s^n} \tag{8.42a}$$

and

$$f(s) = f(0), \tag{8.42b}$$

for $a_1, a_2, \ldots, a_n > 0$ it is evident from

$$A_v(s) = \frac{a(0)}{[1 + a(0)f(0)] + a_1 s + a_2 s^2 + \cdots + a_n s^n} \tag{8.43}$$

that the locations of the natural frequencies of $A_v(s)$ depend on $T(0) = a(0)f(0)$ and that an amplifier with $n \geq 3$ can have one or more poles located in the right half plane, if $T(0)$ becomes sufficiently large.

For the local emitter and collector feedback iterative amplifier chain in Figure 8.5, (8.40) indicates that

$$A_v(s) = \frac{A_{v0}}{1 + s/\omega_c}. \tag{8.44}$$

The natural frequency of $A_v(s)$ is located on the negative real axis at $s = -\omega_c$ so that the amplifier chain is unconditionally stable.

As indicated in Section 8.1, it appears that in a majority of micropower applications purely local feedback—in a multistage amplifier design quite compatible with the goal of minimum quiescent power—provides sufficient gain desensitivity. Fortunately the problems of amplifier stability are minimal in such instances. As multistage feedback loops are employed to improve desensitivity further, higher order terms appear in the denominator of (8.42a) and the danger of self-oscillation increases.

8.5 QUIESCENT POWER

Where bandwidth or power output is not a primary design constraint of a micropower feedback amplifier, quiescent power can be reduced to extremely low values as discussed in Section 3.2 for intermediate stages of low-frequency amplifiers. In those instances in which bandwidth is of major importance its influence on the required quiescent power follows the general pattern discussed in Section 4.2 for wideband amplifiers; for example, (4.1), (4.18), and (8.40) yield

$$P_{DC}^* = 1.2\gamma V_{CC}\omega_c^* C_c n^{3/2} (A_{v0}^*)^{1/n} \left\{ \frac{C_{be}}{C_c} + (1 + g_m R_e)[1 + (A_{v0}^*)^{1/n}] \right\}, \quad (8.45)$$

the total quiescent power for the local emitter and collector feedback amplifier of Figure 8.5. From (8.45) it is evident that a quiescent power penalty arises as a result of emitter feedback as reflected in the factor $(1 + g_m R_e)$. In fact, as the emitter feedback increases so that

$$(1 + g_m R_e)[1 + (A_{v0}^*)^{1/n}] \simeq g_m R_e (A_{v0}^*)^{1/n} \gg C_{be}/C_c, \quad (8.46)$$

it is evident that P_{DC}^* becomes directly proportional to R_e.

REFERENCES

[1] J. G. Linvill and J. F. Gibbons, *Transistors and Active Circuits*, McGraw-Hill, New York, 1961, Chapter 20.
[2] D. Dewitt and A. L. Rossoff, *Transistor Electronics*, McGraw-Hill, New York, 1957, Chapter 8.
[3] M. S. Ghausi, *Principles and Design of Linear Active Circuits*, McGraw-Hill, New York, 1965, Chapter 14.
[4] R. D. Thornton et al., Semiconductor Electronics Education Committee, Vol. 5, *Multistage Transistor Circuits*, Wiley, New York, 1965, Chapters 3, 4.
[5] E. M. Cherry, "An Engineering Approach to the Design of Transistor Feedback Amplifiers," *J. Brit. Instn. Radio Engrs.*, **25**, 127–144 (February 1963).
[6] S. S. Shamis and A. Chertok, "Gain and Sensitivity as Design Criteria for Iterative Transistor Amplifiers," *AIEE Trans.*, Pt. I, *Commun. Electron.*, **80**, 403–407 (July 1961).
[7] J. W. Baker, "Achieving Stable High-Gain Video Amplification Using the Miller-Effect Transformation," *Proc. IEEE*, **51** (No. 6), 911–916 (June 1963).
[8] J. H. Mulligan, "Amplitude Distortion in Transistor Feedback Amplifiers," *AIEE Trans.*, Pt. I, *Commun. Electron.*, **80**, 326–335 (July 1961).

Chapter 9

Harmonic Oscillators

The design constraints for harmonic oscillators [1–3] may include frequency of oscillation, load conductance, power output, output waveform, frequency stability, and amplitude stability as well as other requirements. In most instances, the principal factors that determine the quiescent power of a micropower oscillator are frequency of oscillation, load conductance, and power output. The problem of interest in this chapter is the design of micropower oscillators for minimum quiescent power drain.

9.1 SIMPLIFIED DESIGN

The circuit diagrams for two Hartley oscillators are shown in Figure 9.1. These two circuits will be viewed as common-emitter (Figures 9.1a, 9.1b) and common-base (Figures 9.1c, 9.1d) oscillators in order to facilitate a two-port rather than a three-terminal network analysis [4–5]. In the equivalent circuits in Figure 9.1, the simplified hybrid pi transistor model in Figure 1.6 and an ideal feedback transformer are assumed. The simplified transistor model restricts the analysis to frequencies where

$$r_b \ll \left| \frac{r_{be}}{1 + j\omega r_{be}(C_{be} + C_c)} \right| \tag{9.1}$$

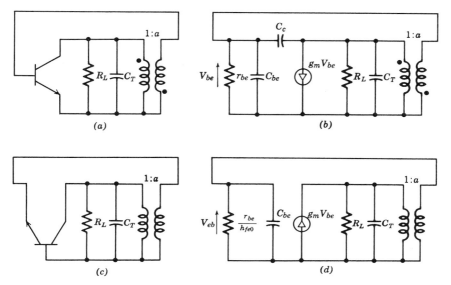

Figure 9.1 Circuit diagrams for common-emitter and common-base harmonic oscillators: (a) CE ac schematic diagram; (b) CE equivalent circuit; (c) CB ac schematic diagram; (d) CB equivalent circuit.

Common-Emitter Hartley Oscillator

If the conductive and susceptive components of Y_0, the oscillator admittance presented to the load, are written as

$$Y_0 = G_0 + jB_0, \tag{9.2}$$

continuous oscillations exist if

$$G_0 + G_L < 0 \tag{9.3a}$$

at such a frequency that

$$B_0 + B_T = 0, \tag{9.3b}$$

where B_T is the susceptance of the tuned circuit in parallel with $G_L = 1/R_L$ [6,7].

From Figure 9.1b, we may calculate

$$G_0 = -a[1 - (a/h_{fe0})]g_m, \tag{9.4}$$

the output conductance of the oscillator. From (9.4), with g_m constant, $\partial G_0/\partial a$ indicates that for an optimum value of the feedback transformer turns ratio

$$a = \frac{h_{fe0}}{2}, \tag{9.5}$$

the maximum value of the negative output conductance

$$G_0 = -\tfrac{1}{4} h_{feo} g_m \tag{9.6}$$

is achieved. Consequently for $a = h_{feo}/2$ the oscillator design is optimized for maximum potential instability.

From (9.3a) and (9.4), it is evident that the limiting value for the load conductance is

$$G_L = -G_0 = a\left(1 - \frac{a}{h_{feo}}\right) g_m. \tag{9.7}$$

If G_L is fixed, substituting (9.7) gives

$$P_{DC}^* = V_{CC} I_C \tag{9.8a}$$

$$= \gamma V_{CC} \frac{G_L}{a(1 - a/h_{feo})} \tag{9.8b}$$

as the quiescent power of the oscillator. From (9.8b), with G_L constant, $\partial P_{DC}^*/\partial a$ indicates that for an optimum turns ratio

$$a = \frac{h_{feo}}{2} \tag{9.9}$$

the quiescent power of the oscillator reaches a minimum value

$$P_{DC}^* = 4\gamma V_{CC} \frac{G_L}{h_{feo}}. \tag{9.10}$$

On the basis of (9.5), (9.7), and (9.9) it is clear that an oscillator design that is optimized for maximum potential instability or maximum negative output conductance for a given quiescent point also is optimized for minimum quiescent power for a given load [8]. Figure 9.2 illustrates a normalized plot of (9.8b) that indicates the dependence of quiescent power on the design of the feedback transformer.

The total ac collector load conductance presented to the transistor is

$$G_c = a^2 g_{be} + G_L. \tag{9.11}$$

For a typical micropower oscillator design, the amplitude of the output signal is limited by emitter cutoff as illustrated in Figure 9.3. Consequently the amplitude of the output voltage is

$$V_{ce} = \frac{I_{CQ}}{G_c}. \tag{9.12}$$

Substituting (9.7) and (9.11) into (9.12) gives

$$V_{ce} = \frac{kT}{q} \frac{1}{a} = \frac{\gamma}{a}, \tag{9.13}$$

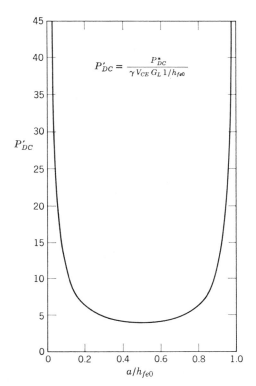

Figure 9.2 Normalized quiescent power P'_{DC} versus feedback transformer turns ratio divided by common-emitter current gain, a/h_{fe0}, for CE oscillator.

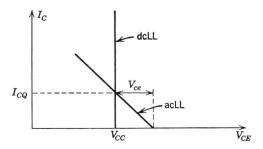

Figure 9.3 Load line diagram for a common-emitter Hartley oscillator.

from which we may derive

$$P_L = \tfrac{1}{2}V_{ce}^2 G_L \qquad (9.14a)$$

$$= \tfrac{1}{2}\gamma \frac{(1 - a/h_{feo})}{a} I_{CQ} \qquad (9.14b)$$

$$\simeq \tfrac{1}{2}\gamma \frac{1}{a} I_{CQ} \quad \text{for} \quad \frac{a}{h_{feo}} \ll 1, \qquad (9.14c)$$

the power delivered to the load, and

$$\eta = \frac{P_L}{P_{DC}^*} = \frac{1}{2}\frac{\gamma}{V_{CC}}\frac{(1 - a/h_{feo})}{a} \qquad (9.15a)$$

$$\simeq \frac{1}{2}\frac{\gamma}{V_{CC}}\frac{1}{a} \quad \text{for} \quad a/h_{feo} \ll 1, \qquad (9.15b)$$

the efficiency.

From Figure 9.3 it is evident that when $V_{ce} = V_{CC}$ the output signal swing is symmetrically limited by emitter cut off and collector saturation. Using (9.13), this requires $a = \gamma/V_{CC} \ll 1$, for which $\eta = \tfrac{1}{2}$. Here it is clear that the design for maximum efficiency is markedly different from the design for maximum potential instability for the circuits in Figure 9.1. For an oscillator designed for maximum potential instability it is evident that the voltages $V_{ce} = (kT/q)(2/h_{feo})$ and $V_{be} = aV_{ce} = kT/q$ are much less than V_{CC}.

On the basis of (9.3b),

$$\omega a^2 \left[C_{be} + \left(\frac{1+a}{a}\right) C_c \right] + \omega C_T - \frac{1}{\omega L_T} = 0 \qquad (9.16)$$

yields the frequency of oscillation

$$\omega_0^2 \simeq \frac{1}{L_T[C_T + a^2(C_{be} + C_c)]}, \qquad (9.17)$$

which is independent of the quiescent or load power for the simple equivalent circuits in Figure 9.1.

Common-Base Hartley Oscillator

Following the pattern of the analysis of the common-emitter oscillator, the load conductance of the common-base oscillator in Figure 9.1c, d can be written as

$$G_L = -G_0 \simeq a(1 - a)g_m, \qquad (9.18)$$

so that the quiescent power is

$$P_{DC}^* = V_{CC}I_C \tag{9.19a}$$

$$= \gamma V_{CC} \frac{G_L}{a(1-a)} \tag{9.19b}$$

and $\partial P_{DC}^*/\partial a$ yields the optimum turns ratio

$$a = \tfrac{1}{2} \tag{9.20}$$

for minimum quiescent power

$$P_{DC}^* = 4\gamma V_{CC}G_L. \tag{9.21}$$

The total ac collector load conductance presented to the transistor is

$$G_c = a^2 g_m + G_L, \tag{9.22}$$

so that the output signal amplitude is

$$V_{cb} = \frac{I_{CQ}}{G_L} = \frac{\gamma}{a}, \tag{9.23}$$

the power output is

$$P_L = \tfrac{1}{2} V_{cb}^2 G_L = \tfrac{1}{2}\gamma \frac{(1-a)}{a} I_{CQ}, \tag{9.24}$$

and the efficiency is

$$\eta = \frac{P_L}{P_{DC}^*} = \frac{1}{2} \frac{\gamma}{V_{CC}} \frac{1-a}{a}. \tag{9.25}$$

The frequency of oscillation is

$$\omega_0^2 \simeq \frac{1}{L_T(C_T + C_c + a^2 C_{be})}. \tag{9.26}$$

Comparing (9.14c) and (9.24),

$$\frac{P_L(CE)}{P_L(CB)} = \frac{\tfrac{1}{2}\gamma[1/a(CE)]I_{CQ}(CE)}{\tfrac{1}{2}\gamma[1/a(CB)]I_{CQ}(CB)} [1 - a(CB)]^{-1}, \tag{9.27}$$

it is clear that for equal load power, $P_L(CE) = P_L(CB)$, and equal quiescent power, $I_{CQ}(CE) = I_{CQ}(CB)$, in the two circuits, the turns ratios must satisfy the condition

$$a(CE) \simeq \frac{a(CB)}{1 - a(CB)}. \tag{9.28}$$

9.2 GENERALIZED DESIGN

Figure 9.4 illustrates a more generalized oscillator circuit configuration in which the transistor is represented by its two-port h parameters and the passive feedback network by an ideal transformer and the impedance Z_F [6,7]. For a very stable frequency of oscillation, a quartz crystal can be included in the feedback network within Z_F. When a design for the transformer feedback arrangement in Figure 9.4 is obtained, other equivalent feedback networks, such as a pi or a tee network, can be calculated readily. The output admittance of the circuit is

$$Y_0 = h_{22} + \frac{(a - h_{12})(h_{21} + a)}{h_{11} + Z_F}. \tag{9.29}$$

From (9.29), we may write

$$Y_0 = h_{22} - Y_n = h_{22} - (G_n + jB_n), \tag{9.30}$$

where

$$G_n = \frac{-a^2 + a[h_{21r} - h_{21r} + \theta(h_{12j} - h_{21j})] + \theta(h_{12r}h_{21j} + h_{21r}h_{12j}) + h_{12r}h_{21r} - h_{12j}h_{21j}}{(h_{11r} + R_F)(1 + \theta^2)} \tag{9.31a}$$

and

$$B_n = \frac{\theta a^2 + a[h_{12j} - h_{21j} - \theta(h_{12r} - h_{21r})] - \theta(h_{12r}h_{21r} + h_{12j}h_{21j}) + h_{12r}h_{21j} + h_{21r}h_{12j}}{(h_{11r} + R_F)(1 + \theta^2)} \tag{9.31b}$$

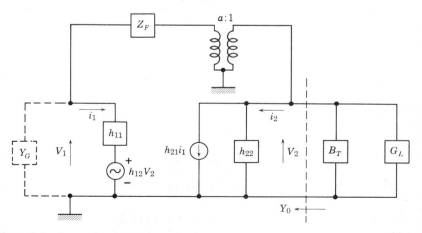

Figure 9.4 Generalized circuit diagram for transistor harmonic oscillator with passive feedback network.

with $h_{ij} = h_{ijr} + jh_{ijj}$,

$$\theta = \frac{h_{11j} + X_F}{h_{11r} + R_F}, \qquad (9.31c)$$

and $Z_F = R_F + jX_F$.

From $\partial G_n/\partial a = 0$ and $\partial G_n/\partial \theta = 0$, the optimum values

$$a_0 = \frac{1}{2}\left[h_{12r} - h_{21r} + \frac{h_{12j} + h_{21j}}{h_{12r} + h_{21r}}(h_{12j} - h_{21j})\right] \qquad (9.32)$$

and

$$\theta_0 = \frac{h_{12j} + h_{21j}}{h_{12r} + h_{21r}} \qquad (9.33)$$

substituted into (9.31a) and (9.31b) yield the maximum value of the negative output conductance

$$G_0(\max) = h_{22r} - \frac{(h_{12r} + h_{21r})^2 + (h_{12j} - h_{21j})^2}{4(h_{11r} + R_F)} \qquad (9.34)$$

and the corresponding output susceptance

$$B_0 = h_{22j} + \theta_0[G_0(\max) - h_{22r}] \qquad (9.35)$$

for $Y_0 = G_0 + jB_0$.

For $Z_F = jX_F$ and an input admittance $Y_G = G_G + jB_G$, (9.34) and (9.35) can be expressed as follows, in terms of the four-pole y parameters:

$$G_0(\max) = g_{22} - \frac{(g_{21} + g_{12})^2 + (b_{21} - b_{12})^2}{4(g_{11} + G_G)} \qquad (9.36)$$

and

$$B_0 = b_{22} - \left(\frac{g_{21} - g_{12}}{b_{21} - b_{12}}\right)[G_0(\max) - g_{22}], \qquad (9.37)$$

where $y_{ij} = g_{ij} + jb_{ij}$.

At relatively low frequencies where (9.1) is valid, (9.36) yields

$$G_0(\max) \simeq -\tfrac{1}{4}h_{fe0}g_m$$

for $g_{22}, g_{12}, b_{21}, b_{12}, G_G = 0$, $g_{21} = g_m$, and $g_{11} = g_m/h_{fe0}$, which corresponds to (9.6).

At relatively high frequencies for which

$$r_b \gg \left|\frac{r_{be}}{1 + j\omega r_{be}(C_{be} + C_c)}\right| \simeq \left|\frac{1}{j\omega(C_{be} + C_c)}\right|, \qquad (9.38)$$

142   Harmonic Oscillators

(9.36) gives

$$G_0(\max) = g_m \frac{C_c}{C_{be} + C_c} - \frac{1}{4r_b}\left(\frac{g_m}{C_{be} + C_c}\right)^2 \frac{1}{\omega_0^2} \qquad (9.39)$$

for

$y_{11} \simeq 1/r_b$, $y_{12} \simeq 0$, $y_{21} \simeq [j\omega_0(C_{be} + C_c)]^{-1}g_m/r_b$, $y_{22} \simeq g_m C_c/(C_{be} + C_c)$
$+ j\omega_0[C_{be}C_c/(C_{be} + C_c)]$, $G_G = 0$,

and ω_0, the frequency of oscillation.

With $G_0 = -G_L$ from (9.3a), substituting (9.39) gives

$$P_{DC}^* = V_{CC}I_C \qquad (9.40a)$$

$$P_{DC}^* = V_{CC}[\tfrac{1}{2}\gamma(C_{be} + C_c)(4r_bC_c\omega_0^2 + \sqrt{(4r_bC_c\omega_0^2)^2 + 16r_bG_L\omega_0^2})], \qquad (9.40b)$$

the minimum quiescent power required by the oscillator at relatively high frequencies for a fixed load conductance G_L.

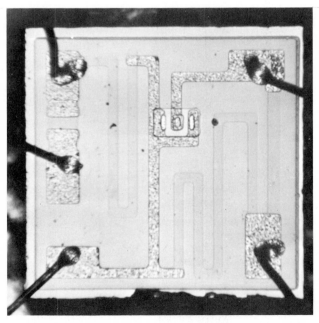

Figure 9.5 Photomicrograph of the monolithic integrated circuit used in the first local oscillator of Figure 5.7. Nichrome thin film resistor line width is 1 mil, chip is 25 × 25 mils (courtesy of U.S. Army Electronics Command [9]).

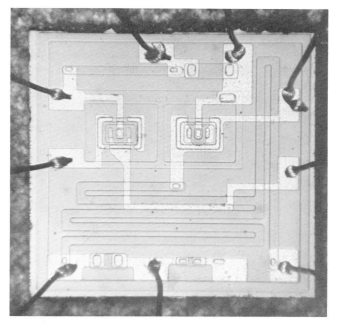

Figure 9.6 Photomicrograph of the monolithic integrated circuit used in the second local oscillator of Figure 5.7. Diffused resistor line width is 1 mil, chip is 40 × 40 mils (courtesy of U.S. Army Electronics Command [9]).

The maximum frequency of oscillation, $\omega_0(\max)$ occurs when $G_0(\max) = G_L = 0$ or for

$$4g_{11}g_{22} = (g_{21} + g_{12})^2 + (b_{21} - b_{12})^2 \qquad (9.41)$$

from (9.36). From (9.39),

$$\omega_0^2(\max) = \frac{\omega_T}{4r_b C_c} \qquad (9.42a)$$

or

$$f_0(\max) = \left(\frac{f_T}{8\pi r_b C_c}\right)^{1/2} \qquad (9.42b)$$

9.3 MICROPOWER MONOLITHIC OSCILLATORS

Figure 5.7 illustrates the schematic diagrams of two micropower local oscillators that utilize monolithic integrated circuits as well as quartz crystals for frequency stability [9]. The dashed rectangles enclose the monolithic integrated circuits, and Figures 9.5 and 9.6 are photomicrographs of the actual silicon chips in which the circuits are fabricated. The salient characteristics of the local oscillators are listed:

	First Oscillator	Second Oscillator
frequency of oscillation	40.3 MHz	10.245 MHz
frequency stability (-30 to $+60°C$)	± 25 ppm/°C	± 25 ppm/°C
mixer injection voltage	100 mV	100–200 mV
quiescent power drain	1.5 mW	900 μW
nominal supply voltage	3.0 V	3.0 V

REFERENCES

[1] L. P. Hunter, *Handbook of Semiconductor Electronics*, 2nd ed., McGraw-Hill, New York, 1962, Section 14.

[2] W. W. Gartner, *Transistors: Principles, Design and Applications*, Van Nostrand, Princeton, N.J., 1960, Chapter 17.

[3] M. V. Joyce and K. K. Clarke, *Transistor Circuit Analysis*, Addison-Wesley, Reading Mass., 1961, Chapter 16.

[4] R. L. Pritchard, "Discussion of Matrix Analysis of Transistor Oscillators," *IRE Trans. Circuit Theory*, **8**, 169–170 (June 1961).

[5] R. Spence, "A Theory of Maximally Loaded Oscillators," *IEEE Trans. Circuit Theory*, **13**, 226–230 (June 1966).

[6] D. F. Page and A. R. Boothroyd, "Instability in Two-port Active Networks," *IRE Trans. Circuit Theory*, **5**, 133–139 (June 1958).

[7] D. F. Page, "A Design Basis for Junction Transistor Oscillator Circuit," *Proc. IRE*, **46**, 1271–1280 (June 1958).

[8] J. D. Meindl and P. H. Hudson, "Low Power Linear Circuits," *IEEE J. Solid-State Circuits*, **1** (No. 2), 100–111 (December 1966).

[9] J. Feit and R. McGinnis, "Small Signal Functional Circuit Units," USAECOM, Fort Monmouth, N.J., Contract No. DA 28-043 AMC-00151(E), Final Report, July 1964–June 1965.

Chapter 10

Direct Current Amplifiers

Direct current or dc amplifiers must respond to essentially unidirectional low-level signals. The primary design constraints that are peculiar to such amplifiers frequently include dc gain, common-mode rejection, input offset voltage, and drift as well as other requirements. The principal purpose of this chapter is to discuss the impact of micropower operation on the performance of dc amplifiers [1,2].

10.1 DC GAIN

Symmetrical or balanced circuit configurations, such as that illustrated in Figure 10.1, are extremely useful in achieving the performance objectives of dc amplifiers [1–4]. For a perfectly balanced amplifier or one in which all homologous parameters are equal in value, an analysis of the incremental circuit model of Figure 10.2 (or Figure 10.3a) yields

$$A_{dd} = -\frac{v_{c2} - v_{c1}}{v_{s2} - v_{s1}} \simeq \frac{R_C R_L/(R_C + R_L)}{1/g_m + R_E + R_S/h_{fe0}}, \qquad (10.1a)$$

the low-frequency or incremental dc differential voltage gain of the amplifier with $R_{S1} = R_{S2} = R_S$, $R_{E1} = R_{E2} = R_E$, $R_{C1} = R_{C2} = R_C$, $h_{fe01} = h_{fe02} = h_{fe0}$, and $g_{m1} = g_{m2} = g_m = qI_C/kT$. Typically, the value of A_{dd} is not influenced significantly by the quiescent current or power drain of the circuit since R_C, R_L, and R_E tend to change in direct proportion to $1/g_m$.

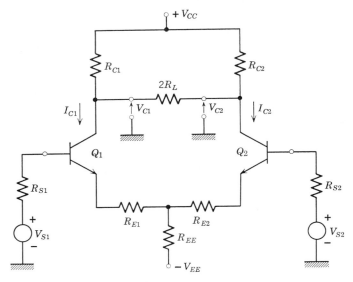

Figure 10.1 Schematic diagram of simple symmetrical dc amplifier circuit.

For a perfectly balanced amplifier, an analysis of the incremental model of Figure 10.2 (or Figure 10.3b) yields

$$A_{cc} = -\frac{v_{c2} + v_{c1}}{v_{s2} + v_{s1}} \simeq \frac{R_C}{1/g_m + R_E + R_S/h_{fe0} + 2R_{EE}} \quad (10.1b)$$

the common mode voltage gain of the amplifier.

Since the single ended output voltages are

$$v_{c1} = -A_{cc}\left(\frac{v_{s1} + v_{s2}}{2}\right) - A_{dd}\left(\frac{v_{s1} - v_{s2}}{2}\right)$$

and $\qquad\qquad(10.1c)$

$$v_{c2} = -A_{cc}\left(\frac{v_{s1} + v_{s2}}{2}\right) + A_{dd}\left(\frac{v_{s1} - v_{s2}}{2}\right)$$

it is clear that unless $A_{cc} \ll A_{dd}$ a small differential mode input signal $\frac{1}{2}(v_{s1} - v_{s2})$ could be swamped at the output by a large common mode input $\frac{1}{2}(v_{s1} + v_{s2})$. Consequently, a large discrimination factor

$$F = \frac{A_{dd}}{A_{cc}} = \frac{R_L}{R_C + R_L} \frac{1/g_m + R_E + R_S/h_{fe0} + 2R_{EE}}{1/g_m + R_E + R_S/h_{fe0}} \quad (10.1d)$$

is necessary. It is evident that the increase in R_{EE} which can accompany a reduction in quiescent current drain is useful in increasing the discrimination factor.

10.2 COMMON-MODE REJECTION

A particular virtue of the symmetrical amplifier in Figure 10.1 is that it very effectively rejects a signal that appears simultaneously at both input terminals. The degree of rejection of such unwanted common-mode signals depends on the degree of symmetry or balance of the circuit. For an incremental common-mode signal $v_{s1} = v_{s2} = v_s$, an analysis based on the circuit model in Figure 10.2 yields the common mode-to-differential mode transfer gain

$$A_{dc} = -\frac{v_{c2} - v_{c1}}{2v_s} \qquad (10.2a)$$

$$A_{dc} \simeq \frac{R_{P1}(R_{S2}/h_{fe02} + 1/g_{m2} + R_{E2}) - R_{P2}(R_{S1}/h_{fe01} + 1/g_{m1} + R_{E1})}{2R_{EE}[(R_{S2}/h_{fe02} + 1/g_{m2} + R_{E2}) + (R_{S1}/h_{fe01} + 1/g_{m1} + R_{E1})]},$$

$$(10.2b)$$

for $R_{P1} = R_{C1}R_L/(R_{C1} + R_L)$, $R_{P2} = R_{C2}R_L/(R_{C2} + R_L)$, and $R_S/h_{fe0}R_{EE} \ll 1$, as the ratio of the differential-mode output voltage to the common-mode input voltage for an imperfectly balanced circuit.

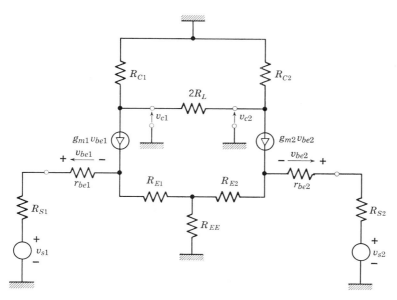

Figure 10.2 Incremental equivalent circuit of a symmetrical dc amplifier.

The output voltage described by (10.2) represents undesirable cross-coupling in comparison to the desired output signal represented by (10.1a) and must be suppressed. A useful measure of the relative rejection of unwanted common-mode signals is the common-mode rejection ratio

$$CMRR = \frac{A_{dd}}{A_{dc}}$$

$$\simeq \frac{4R_{EE}}{(R_{S2}/h_{fe02} - R_{S1}/h_{fe01}) + (1/g_{m2} - 1/g_{m1}) + (R_{E2} - R_{E1})} \quad (10.3)$$

for $R_{P1} \simeq R_{P2} = R_P$. For a perfectly balanced circuit, (10.3) indicates that $CMRR \to \infty$. In most practical cases a large $CMRR$ is achieved for small imbalances by increasing the effective value of R_{EE}. Typically

$$V_{EE} \simeq (I_{E1} + I_{E2})R_{EE} \simeq 2I_C R_{EE} \quad (10.4)$$

for $I_{C1} \simeq I_{C2} = I_C$ so that with a fixed supply voltage, V_{EE}, reducing I_C to the micropower range permits R_{EE} to increase. This increases the $CMRR$,

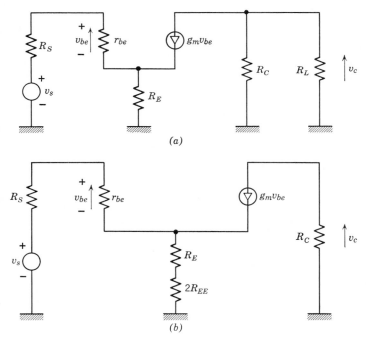

Figure 10.3 (a) Differential mode incremental equivalent circuit of a perfectly balanced symmetrical dc amplifier; (b) common mode incremental equivalent circuit of a perfectly balanced symmetrical dc amplifier.

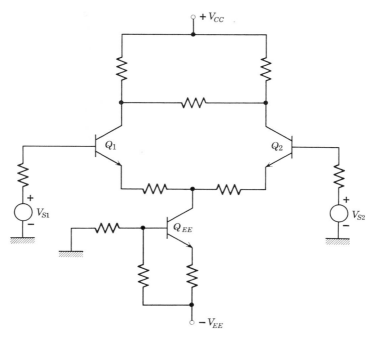

Figure 10.4 Symmetrical dc amplifier with improved common-mode rejection.

in the typical case where $(R_{S2}/h_{fe02} - R_{S1}/h_{fe01}) > (1/g_{m2} - 1/g_{m1})$ and $R_{E2} = R_{E1} = 0$, in direct proportion to R_{EE}.

The effective value of R_{EE} can be increased markedly by employing the dynamic collector output impedance of a transistor as illustrated in Figure 10.4. The low-frequency model in Figure 10.5, for which

$$\frac{1}{r_c} = \frac{1}{2}\left(\frac{1}{W}\frac{dW}{dV_{CB}}\right)\left(\frac{W}{L_b}\right)^2 I_C, \tag{10.5a}$$

$$\frac{1}{r_{ce}} = \left(\frac{1}{W}\frac{dW}{dV_{CB}}\right) I_C \tag{10.5b}$$

and W is the electrical base width of the transistor illustrates the further possibility of reducing I_C to increase the transistor dynamic impedances themselves. Common-mode feedback to the base of the transistor Q_{EE} in Figure 10.4 as well as the use of a cascode pair in place of both Q_1 and Q_2 in Figure 10.1 can be employed to improve the CMRR further [5].

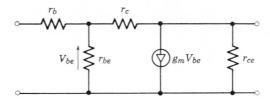

Figure 10.5 Low-frequency small-signal transistor hybrid π model.

10.3 INPUT OFFSET VOLTAGE

The static differential output voltage of the circuit in Figure 10.1 can be written as [4]

$$V_{C2} - V_{C1} \simeq -(V_{S2} - V_{S1})\frac{\alpha_{N1}R_{P1} + \alpha_{N2}R_{P2}}{\Delta}$$

$$+ \frac{(V_{BE2} - V_{BE1})(\alpha_{N1}R_{P1} + \alpha_{N2}R_{P2})}{\Delta}$$

$$+ \frac{(I_{CB01}R_{S1} - I_{CB02}R_{S2})(\alpha_{N1}R_{P1} + \alpha_{N2}R_{P2})}{\Delta}$$

$$+ (I_{CB01}R_{P1} - I_{CB02}R_{P2}) + \frac{V_{EE}}{R_{EE}}$$

$$\times \frac{(R_{S1}/h_{FE1} + R_{E1} - R_{S2}/h_{FE2} - R_{E2})(\alpha_{N1}R_{P1} + \alpha_{N2}R_{P2})}{\Delta},$$

(10.6)

where

$$\Delta = R_{E1} + R_{E2} + \frac{R_{S1}}{h_{FE1}} + \frac{R_{S2}}{h_{FE2}} + \frac{R_{S1}R_{S2}}{h_{FE1}h_{FE2}} R_{EE},$$

$$R_{P1} = \frac{R_{C1}R_L}{R_{C1} + R_L} \quad \text{and} \quad R_{P2} = \frac{R_{C2}R_L}{R_{C2} + R_L},$$

assuming $R_S/h_{FE}R_{EE} \ll 1$ and using the static transistor model of Figure 10.6. In order to achieve zero output voltage, $V_{C2} - V_{C1} = 0$, for zero input voltage, $V_{S2} - V_{S1} = 0$, all homologous elements in the circuit must be equal in value. Assuming I_{CB0} is negligible, for a slight imbalance an input voltage

$$\Delta V_S = V_{S2} - V_{S1}$$

$$\simeq (V_{BE2} - V_{BE1}) + \frac{V_{EE}}{R_{EE}}\left(\frac{R_{S1}}{h_{FE1}} + R_{E1} - \frac{R_{S2}}{h_{FE2}} - R_{E2}\right) \quad (10.7)$$

is required to obtain zero output voltage.

From (10.4) and (10.7) it is clear that decreasing the quiescent current drain of the amplifier reduces the input offset voltage ΔV_S, as well as the associated input offset current. A low source resistance R_S is desirable for a small ΔV_S, which can be reduced to zero at a given temperature and operating point by using an adjustable tapped resistance for R_{E1} and R_{E2}.

10.4 DRIFT

The sensitivity of a low-level dc amplifier is limited by drift or changes in the value of ΔV_S because of unequal temperature coefficients for homologous circuit elements [6–9]. From (10.7),

$$\frac{\partial(\Delta V_S)}{\partial T} = \frac{\partial(V_{BE2} - V_{BE1})}{\partial T} - \frac{V_{EE}}{R_{EE}}\left[\frac{R_{S1}}{h^2_{FE1}}\frac{\partial h_{FE1}}{\partial T} - \frac{R_{S2}}{h^2_{FE2}}\frac{\partial h_{FE2}}{\partial T}\right], \quad (10.8)$$

assuming $\partial V_{EE}/\partial T = \partial R_{EE}/\partial T = \partial R_E/\partial T = \partial R_S/\partial T = 0$, indicates that drift caused by h_{FE} mismatch can be reduced to negligible values by increasing R_{EE}. From (1.13), the relationship between emitter voltage and collector current can be written as

$$I_C = \alpha_N I_{EBS} \epsilon^{qV_{BE}/kT}. \quad (10.9a)$$

Since the emitter junction saturation current I_{EBS} is given by

$$I_{EBS} = CT^r \epsilon^{-q\phi_0/kT}, \quad (10.9b)$$

combining (10.9a) and (10.9b) yields [8,9]

$$I_C = (\alpha_N C)T^r \epsilon^{-q(\phi_0 - V_{BE})/kT}, \quad (10.10)$$

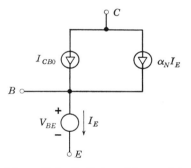

Figure 10.6 Static transistor model for active region.

where α_N, C, r, ϕ_0, q, and k are essentially independent of temperature. Here C is a constant dependent on transistor area and impurity concentration, r is a constant dependent or minority carrier mobility with a typical value of 1.5 for a silicon transistor, and ϕ_0 is the semiconductor energy band gap at $T = 0°K$ ($\phi_0 = 1.205$ V for Si).

Solving (10.10) for V_{BE} gives

$$V_{BE} = \phi_0 + \frac{kT}{q}(\ln I_C - \ln \alpha_N - \ln C - r \ln T). \quad (10.11)$$

Thus

$$\left.\frac{\partial V_{BE}}{\partial T}\right|_{I_C} = \frac{k}{q}(\ln I_C - \ln \alpha_N - \ln C - r - r \ln T) \quad (10.12)$$

or

$$\left.\frac{\partial V_{BE}}{\partial T}\right|_{I_C} = -\frac{kr}{q} - \frac{\phi_0 - V_{BE}}{T}. \quad (10.13)$$

Assuming r and ϕ_0 are exactly equal for two matched transistors, substituting (10.13) into (10.8) gives

$$\frac{\partial(\Delta V_S)}{\partial T} \simeq \frac{V_{BE2} - V_{BE1}}{T}. \quad (10.14)$$

A second assumption on which (10.14) is based is that the junction temperatures of the two transistors are equal. This condition is promoted by the small internal power dissipation accompanying micropower operation.

The important information revealed by (10.14) is that the thermal drift of a dc amplifier can be reduced to extremely small values (i.e., $\partial(\Delta V_S)/\partial T \simeq 0$) if the two transistors are operated at equal base-emitter voltages (i.e., $V_{BE2} = V_{BE1}$). For the condition $V_{BE2} = V_{BE1}$, (10.8) indicates that the collector currents are related by the expression

$$I_{C1} = \frac{\alpha_{N1} I_{EBS1}}{\alpha_{N2} I_{EBS2}} I_{C2} \quad (10.15)$$

and in general are unequal. To maintain $V_{C2} - V_{C1} = 0$ for $V_{S2} - V_{S1} = 0$ and $I_{C1} \neq I_{C2}$, the collector resistors must be adjusted so that

$$R_{C1} = \frac{I_{C2}}{I_{C1}} R_{C2}. \quad (10.16)$$

Drifts of less than 10^{-6} V/°C can be achieved using this technique [8,9].

10.5 NOISE

In dc amplifiers with extremely low drift rates, $1/f$ or flicker noise is a limiting factor to the sensitivity [9,10]. From (6.1), the flicker noise generator used in the hybrid π transistor model is described by

$$\overline{i_f^2} = K_f I_B^\lambda \frac{1}{f^\alpha} \Delta f, \quad (10.17)$$

where K_f, λ, and α are empirical constants of the device. From (10.17), it is evident that a small quiescent current, $I_C = h_{FE} I_B$, tends to reduce flicker noise in a dc amplifier.

The noise figure of an amplifier can be expressed as [10]

$$F - 1 = (F_0 - 1)\left[1 + \left(\frac{f_{cl}}{f}\right)^\alpha\right], \quad (10.18)$$

where F_0 is the midband noise figure and f_{cl} is the lower noise corner frequency. If (10.18) is integrated over the amplifier bandwidth,

$$\overline{e_{eq}^2} = 4kTR_S(F_0 - 1)[\Delta f - \Delta f'] \qquad (10.19)$$

is the amplifier equivalent mean square noise voltage referred to the input for

$$\Delta f = f_h - f_l, \qquad (10.20a)$$

the amplifier bandwidth effective for white noise, and

$$\Delta f' = \int_{f_l}^{f_h} \left(\frac{f_{cl}}{f}\right)^\alpha df, \qquad (10.20b)$$

the effective $1/f$ noise bandwidth with f_h and f_l the upper and lower cutoff frequencies of the amplifier. However, f_l is not a cutoff frequency in the true sense of the word, but the inverse of the total time of continuous operation. For $\alpha > 1$,

$$\Delta f' = \frac{f_{cl}^\alpha}{\alpha - 1}\left[\frac{1}{f_l^{\alpha-1}} - \frac{1}{f_h^{\alpha-1}}\right]. \qquad (10.21)$$

If it is assumed that the noise voltage has a Gaussian distribution, the peak value will be limited to within $\pm 2.6\sqrt{\overline{e_{eq}^2}}$ for 99% of the time. Combining this result with (10.19) through (10.21) yields

$$|E_{eq}| = 5.2\left\{kTR_S(F_0 - 1)\left[f_h - f_l + \frac{f_{cl}^\alpha}{\alpha - 1}\left(\frac{1}{f_l^{\alpha-1}} - \frac{1}{f_h^{\alpha-1}}\right)\right]\right\}^{1/2}, \qquad (10.22)$$

the limiting amplitude of the noise voltage fluctuations for 99% of the time.

Typically, (10.22) indicates (a) that $|E_{eq}|$ can exceed 10^{-6} volts, which is large enough to dominate the sensitivity of a well-stabilized dc amplifier, (b) that $|E_{eq}|$ increases rather slowly with increasing $1/f_l$ or time and very little with increasing f_h, and (c) that $|E_{eq}|$ increases quite rapidly with increasing $\alpha > 1$.

10.6 FREQUENCY RESPONSE

The upper 3 dB cutoff frequency, the roll-off characteristics, and the ac stability of dc amplifiers used as operational amplifiers are of considerable importance. The impact of micropower operation on these properties is discussed in Chapters 4 and 8.

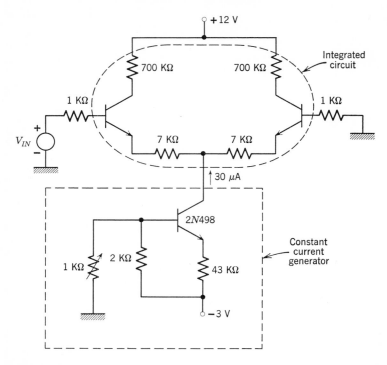

Figure 10.7 Schematic diagram of monolithic micropower dc amplifier (after Bittman et al. [11]).

10.7 A MONOLITHIC MICROPOWER dc AMPLIFIER

A simple micropower dc amplifier that has been fabricated as a monolithic integrated circuit is illustrated in Figure 10.7 [11]. The circuit provides a differential-mode voltage gain of 30 dB, a common-mode gain of -50 dB, and a 3-dB bandwidth of 10 kHz. The micropower transistors exhibit a current gain of 10 at 10^{-9} A and compatible thin film silicon-chromium resistors are used.

10.8 A MICROPOWER OPERATIONAL AMPLIFIER

Figure 10.8 illustrates the schematic diagram of a micropower low-noise monolithic operational amplifier [12]. The input stage consists of an NPN differential pair biased from a temperature compensating current source.

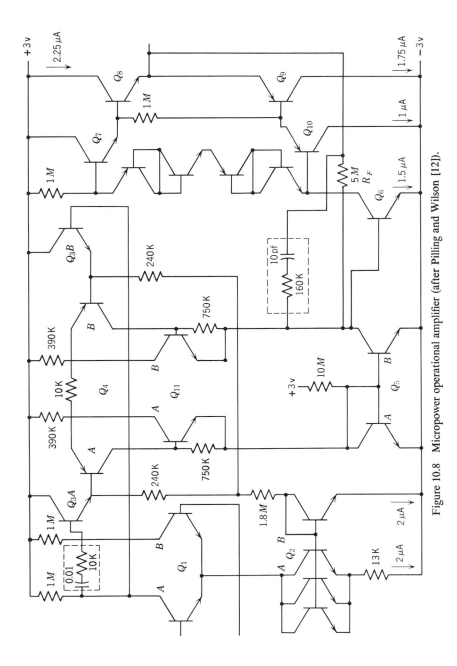

Figure 10.8 Micropower operational amplifier (after Pilling and Wilson [12]).

The second stage is a composite PNP differential stage that level-shifts while providing additional gain. The output stage achieves minimum distortion by using diodes to compensate the dead zone introduced by the output transistor base-emitter voltage drops. It can be operated class AB to minimize quiescent power drain. The transistors in the circuit exhibit minimum current gain fall-off at low currents, and the resistors are compatible silicon-chromium thin film elements. The salient features of the performance are

supply voltage	± 3 V (to ± 15 V)
quiescent power drain	100 μW
output swing ($R_L = 1$ kΩ)	± 1.5 V
dc gain	20,000
input impedance	7 MΩ
output impedance	15 Ω
common-mode rejection	6 μV/volt
power supply rejection	7 μV/volt
temperature range	-55–$125°$C
open-loop frequency cutoff	10 kHz

REFERENCES

[1] R. D. Middlebrook, *Differential Amplifiers*, Wiley, New York, 1964.
[2] R. D. Thornton et al., *Multistage Transistor Circuits*, SEEC Vol. 5, Wiley, New York, 1965, Chapter 6.
[3] D. W. Slaughter, "The Emitter-coupled Differential Amplifier," *IRE Trans. Circuit Theory*, **3**, 51–53 (March 1956).
[4] R. H. Okada, "Stable Transistor Wideband d-c Amplifiers," *Trans. AIEE Commun. Electron.*, **79**, 26–33 (March 1960).
[5] G. Meyer-Brotz and A. Kley, "The Common-Mode Rejection of Transistor Differential Amplifiers," *IEEE Trans. Circuit Theory*, **13**, 171–175 (June 1966).
[6] Hilbiber, "A New DC Transistor Differential Amplifier," *IRE Trans. Circuit Theory*, **CT-8**, 434–439 (December 1961).
[7] P. J. Beneteau et al., "Drift Compensation in dc Amplifiers," *Solid State Design*, **5** 19–23 (May 1964).
[8] D. R. Hilbiber, "Stable Differential Amplifier Designed Without Choppers," *Electronics*, **38**, 73–75 (January 25, 1965).
[9] A. H. Hoffait and R. D. Thornton, "Limitations of Transistor dc Amplifiers," *Proc. IEEE*, **52**, 179–184 (February 1964).
[10] T. C. Verster, "Low Frequency Noise in dc Amplifiers," *Proc. IEEE*, **54**, 1211–1213 (September 1966).
[11] C. A. Bittman et al., "Micropower Functional Electronic Bloks," Technical Report, AFAL-TR-66-338, 83–89 (September 1966).
[12] D. J. Pilling and G. H. Wison, "A Micropower Low Noise Integrated Operational Amplifier," Proc. of 1967 International Telemetering Conf., Washington, D.C., **3**, 509-521 (November 1967).

Chapter 11

Bipolar Transistor Digital Circuits

In most digital systems the key circuit is the basic logic gate. It serves as the elemental building block from which complex system functions can be implemented. Frequently the design constraints of a basic gate include the logic function to be performed, fan-in and fan-out, cutoff and saturated stability margins, immunity to both manufacturing and environmental tolerances in the values of circuit elements and to variations in supply voltages, operating speed, and (especially in micropower applications) the average power consumption of the circuit. The principal purpose of this chapter is to consider the design of bipolar transistor gate circuits for operation in the micropower range [1–7].

11.1 THE BASIC INVERTER CIRCUIT

Perhaps more than any other circuit configuration, the simple transistor inverter, which performs the logical operation of complementation, exemplifies the salient features of the behavior of digital circuits. The static and dynamic performance of a single micropower inverter is discussed in this section.

Static Behavior [1,2]

The schematic diagram of a simple inverter circuit with an equivalent load resistance R_L is illustrated in Figure 11.1. The input loop equation for this circuit is

$$V_B = I_B R_B + V_{BE}, \qquad (11.1a)$$

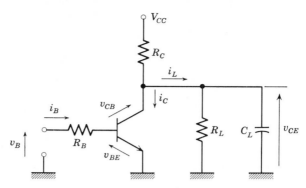

Figure 11.1 Transistor inverter circuit.

and the input characteristic of the transistor is given by

$$I_B = (1 - \alpha_N)I_{EBS}(\epsilon^{V_{BE}/\gamma} - 1) + (1 - \alpha_I)I_{CBS}(\epsilon^{-V_{CB}/n\gamma} - 1) \quad (11.1b)$$

from (1.12) with $\gamma = kT/q$. The output circuit equations are

$$V_{CC} = (I_C + I_L)R_C + V_{CE}$$

and

$$V_{CE} = V_{CB} + V_{BE} = I_L R_L, \quad (11.1c)$$

and the output characteristic equation of the transistor is

$$I_C = \alpha_N I_{EBS}(\epsilon^{V_{EB}/\gamma} - 1) - I_{CBS}(\epsilon^{-V_{CB}/n\gamma} - 1). \quad (11.1d)$$

A series of four graphical solutions to (11.1) corresponding to four different input voltages, V_{B1} through V_{B4}, is illustrated in Figure 11.2. From these solutions the forward voltage transfer characteristic of the inverter can be constructed as shown in Figure 11.3.

When the transistor is operating in the cutoff region,

$$V_B \simeq V_{BE} < V_{B2} \quad (11.2a)$$

$$I_B \simeq 0 \quad (11.2b)$$

$$V_{CE} \simeq V_{CC} \frac{R_L}{R_L + R_C} \quad (11.2c)$$

and

$$I_C \simeq 0 \quad (11.2d)$$

from (11.1). In the active region,

$$V_B = I_B R_B + V_{BE} \quad \text{for} \quad V_{B2} \leq V_B \leq V_{B3}, \quad (11.3a)$$

$$I_B \simeq (1 - \alpha_N)I_{EBS}\epsilon^{V_{BE}/n\gamma} \quad (11.3b)$$

$$V_{CE} = \frac{R_L}{R_L + R_C}(V_{CC} - I_C R_C) \quad (11.3c)$$

and
$$I_C \simeq \alpha_N I_{EBS} \epsilon^{V_{BE}/\gamma}. \tag{11.3d}$$

Assuming $V_{BE} \simeq V_{B2}$, the relationships $I_B = (V_B - V_{B2})/R_B$, $I_C = \alpha_N/(1 - \alpha_N)I_B \simeq h_{FE}I_B$, and (11.3c) provide a simple yet accurate description of the transfer characteristic for the active region. In the saturation region, $V_B > V_{B3}$, (11.1) applies. Assuming V_{BE} is taken at its final value

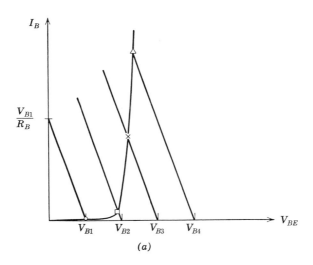

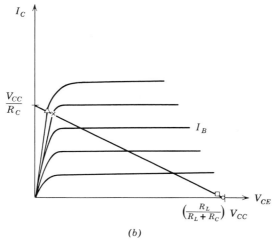

Figure 11.2 (a) Input loadline diagram of inverter; (b) output loadline diagram of inverter.

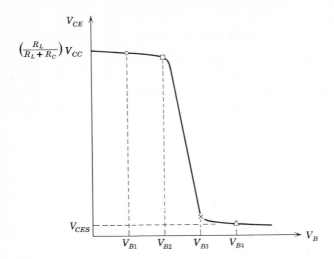

Figure 11.3 Forward voltage transfer characteristic of an inverter.

in the active region, (11.1a) gives I_B, (11.1c) gives

$$I_{CS} \simeq \frac{V_{CC}}{R_C} \quad \text{for} \quad V_{CC} \gg V_{CES}, \tag{11.4}$$

and then (11.1b) and (11.1d) solved simultaneously yield the saturation voltage

$$V_{CES} = \gamma \ln \left[\frac{1}{\alpha_I} \frac{1 + (1 - \alpha_I)I_C/I_B}{1 - \frac{1}{h_{FE}} I_C/I_B} \right] \tag{11.5}$$

for $V_B > V_{B3}$ to complete the simplified quantitative description of the static inverter transfer function in Figure 11.3.

For a fixed supply voltage, the selection of the saturated collector current I_{CS} sets the value $R_C \simeq V_{CC}/I_{CS}$ from (11.4) and the range of values

$$R_B \leq \frac{V_B - V_{BES}}{I_{CS}/h_{FE}} \tag{11.6}$$

from (11.1a). Consequently the average power consumption of the inverter is

$$P_{DC}^* = \frac{1}{2} \left[V_{CC}^2 \left(\frac{1}{R_C} + \frac{1}{R_C + R_L} \right) + V_B \frac{(V_B - V_{BES})}{R_B} \right] \tag{11.7a}$$

or, typically,

$$P_{DC}^* \simeq \tfrac{1}{2} V_{CC}^2 \frac{1}{R_C} \simeq \tfrac{1}{2} V_{CC} I_{CS}. \tag{11.7b}$$

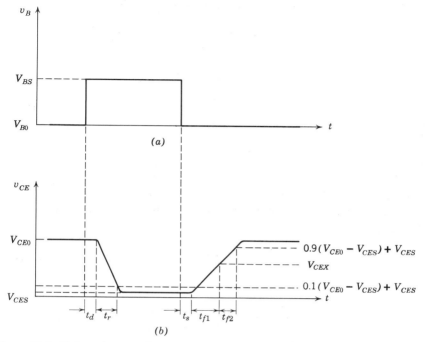

Figure 11.4 Voltage input and output waveforms for the simple transistor inverter of Figure 11.1.

Dynamic Behavior [1,2,8]

The transient response of the output voltage of the inverter, $v_{CE}(t)$, to a step change in the input voltage $v_B(t)$ is illustrated in Figure 11.4. For comparison with Figure 11.3, we may assume $V_{B0} < V_{B1}$, $V_{BS} > V_{B3}$, and $V_{CE0} = R_L/(R_L + R_C)V_{CC}$. Except for the dc collector resistance R_C, the load on the inverter is represented by an equivalent parallel resistance R_L and capacitance C_L as shown by Figure 11.1.

During the turn-on delay time t_d [8–11], the transistor can be represented in the circuit in Figure 11.1 by the cutoff region model in Figure 1.4c. The Kirchhoff equations for the input and output circuits are

$$\frac{v_B - v_{BE}}{R_B} \simeq C_{BE}\frac{dv_{BE}}{dt} + C_{CB}\frac{d}{dt}(v_{BE} - v_{CE}) \qquad (11.8a)$$

and

$$i_C \simeq 0 \qquad (11.8b)$$

from Figure 11.1. The solution of the resultant equation

$$\tau_d \frac{dv_{BE}}{dt} + v_{BE} = V_{BS} \tag{11.8c}$$

yields the approximate value of the delay time

$$t_d = \tau_d \ln \frac{V_{BS} - V_{B0}}{V_{BS} - V_{BET}}, \tag{11.9a}$$

where

$$\tau_d = R_B(C_{BE} + C_{CB}) \tag{11.9b}$$

and the "threshold" voltage of the transistor $V_{BET} \simeq V_{B2}$ in Figure 11.2a and Figure 11.3. As described by (11.9a), during t_d the excursion of $v_{BE}(t)$ is $V_{B0} \leq v_{BE}(t) \leq V_{BET}$. Also the average values used for the transistor capacitances are

$$C_{BE} = \frac{1}{V_{BET} - V_{B0}} \int_{V_{B0}}^{V_{BET}} C_{be}(V_{BE}) \, dV_{BE} \tag{11.10a}$$

and

$$C_{CB} = \frac{1}{(V_{C0} - V_{BET}) - (V_{C0} - V_{B0})} \int_{V_{C0}-V_{B0}}^{V_{C0}-V_{BET}} C_{cb}(V_{CB}) \, dV_{CB} \tag{11.10b}$$

where typical behavior of $C_{be}(V_{BE})$ and $C_{cb}(V_{CB})$ is illustrated in Figure 1.5. For each switching interval the average values of the transistor parameters for that particular interval should be used in calculations pertaining to it.

During the rise time t_r [8,12–18], the active region model in Figure 1.4b can be used to represent the transistor in the circuit in Figure 11.1. The Kirchhoff equations for the input and output circuits are

$$i_B = \frac{i_C}{h_{FE}} + C_{BE} \frac{dv_{BE}}{dt} + C_{CB} \frac{d(v_{BE} - v_{CE})}{dt} \tag{11.11a}$$

and

$$i_C \simeq \frac{V_{CC} - v_{CE}}{R_C} - \frac{v_{CE}}{R_L} - C_L \frac{dv_{CE}}{dt}. \tag{11.11b}$$

The solution of the resultant equation

$$\frac{h_{FE}}{\omega_T} R_{EQ} C_L \frac{d^2 v_{CE}}{dt^2} + \left(\frac{h_{FE}}{\omega_T} + R_{EQ} C_L + h_{FE} R_{EQ} C_{CB}\right) \frac{dv_{CE}}{dt} + v_{CE}$$
$$= \frac{R_{EQ}}{R_C} V_{CC} - h_{FE} R_{EQ} I_{BS} \tag{11.11c}$$

yields the approximate value of the rise time

$$t_r \simeq \tau_r \ln\left(\frac{I_{BS}}{I_{BS} - 0.9 I_{CS}/h_{FE}}\right), \qquad (11.12a)$$

where

$$\tau_r = \frac{h_{FE}}{\omega_T} + R_{EQ}C_L + h_{FE}R_{EQ}C_{CB}, \qquad (11.12b)$$

$$I_{CS} \simeq \frac{V_{CC}}{R_C}, \qquad (11.12c)$$

$$R_{EQ} = \frac{R_C R_L}{R_C + R_L}, \qquad (11.12d)$$

$$I_{BS} \simeq \frac{V_{BS} - V_{BET}}{R_B}, \qquad (11.12e)$$

and

$$\frac{4(h_{FE}/\omega_T)R_{EQ}C_L}{\tau_r^2} \ll 1. \qquad (11.12f)$$

During t_r, the excursion of $v_{CE}(t)$ is $V_{CE0} \geq v_{CE}(t) \geq 0.1(V_{CE0} - V_{CES}) + V_{CES}$. Average values for h_{FE}, ω_T, and C_{CB} for this interval can be used in (11.12). Unless (11.12f) applies, both exponential terms in the solution of (11.11c) must be retained and an explicit expression for t_r cannot be determined.

The transistor model in Figure 1.2b can be used in the calculation of the storage time [11,12,14,16–19]

$$t_s = \tau_s \ln \frac{I_{BS} - I_{B0}}{I_{CS}/h_{FE} - I_{B0}}, \qquad (11.13)$$

where

$$\tau_s = \frac{\omega_{\alpha N} + \omega_{\alpha I}}{\omega_{\alpha N}\omega_{\alpha I}(1 - \alpha_N \alpha_I)}, \qquad (11.14a)$$

and

$$I_{B0} = -\frac{V_{BE3} - V_{B0}}{R_B}. \qquad (11.14b)$$

The forward and reverse common-base α cutoff frequencies are denoted by $\omega_{\alpha N}$ and $\omega_{\alpha I}$, respectively. At these frequencies, $|\alpha_N|$ and $|\alpha_I|$ are diminished to $1/\sqrt{2}$ times their respective low-frequency values. In (11.14b), V_{BE3} is the base-emitter voltage corresponding to V_{B3} as indicated in Figure 11.2a and Figure 11.3.

For the typical micropower transistor, $\alpha_N \simeq 1$ and $\omega_{aN} \gg \omega_{aI}$, so that (11.14a) gives

$$\tau_s \simeq \frac{1}{\omega_{aI}(1 - \alpha_I)}. \qquad (11.15)$$

This result can also be achieved by assuming $dv_{CE}/dt = 0$ when we solve the differential equations that describe the inverter behavior during the storage interval. The boundary condition that marks the duration of t_s for the transistor model in Figure 1.4a is the reduction to zero, essentially, of the forward current through the collector diode of the micropower transistor. As Figure 1.5 indicates, this condition prevails for a positive or forward collector junction voltage. The average value of C_{CB} and the excursion of $v_{CE}(t)$ during the ensuing fall time should reflect this condition. Particularly, when a charge control model of the transistor is employed, the storage time constant τ_s is treated as a basic switching transistor parameter, the apparent lifetime that characterizes the recombination of the stored charge in the base region in excess of the amount present at the saturation edge of the active region.

In general the fall time t_f [11–18] of an inverter operating in the micropower range must be divided into two intervals [8,20]: t_{f1}, during which the transistor operates in the active region, and t_{f2}, during which the transistor is cutoff so that $t_f = t_{f1} + t_{f2}$. The equivalent circuit of the inverter during the active region fall time is identical to the one used to calculate t_r. Thus

$$v_{CE}(t) \simeq (I_{CS} - h_{FE}I_{B0})R_{EQ}(1 - \epsilon^{-t/\tau_r}) \qquad (11.16)$$

for $0 \le t \le t_{f1}$ and $\tau_{f1} = \tau_r$. Substituting (11.17a) into (11.11b) gives

$$i_C(t) \simeq (I_{CS} - h_{FE}I_{B0})\left(1 - \frac{\tau_{f2}}{\tau_r}\right)\epsilon^{-t/\tau_r} + h_{FE}I_{B2} \qquad (11.17)$$

for $0 \le t \le t_{f1}$, where $\tau_{f2} = R_{EQ}C_L$. Now $i_C(t_{f1}) = 0$ yields

$$t_{f1} \simeq \tau_r \ln\left[\frac{I_{B0} - I_{CS}/h_{FE}}{I_{B0}}\left(1 - \frac{\tau_{f2}}{\tau_r}\right)\right]. \qquad (11.18a)$$

Note that (11.18a) may yield $t_{f1} < 0$ as $\tau_{f2} \to \tau_r$. This possibility is a result of using a single exponential term to approximate $v_{CE}(t)$ and $i_C(t)$. Clearly, as $\tau_{f2} \to 0$ (11.18a) yields the usual expressions [11–16] for t_{f1} with $i_C(t_{f1}) = 0$. In virtually all cases of practical interest in micropower circuits, $t_{f1} < 0$ from (11.18a) merely implies a negligible value of t_{f1} for the circuit under consideration. In such instances an approximate value for t_{f1} can be calculated from (11.11a) by assuming $dv_{CE}/dt \simeq 0$ during t_{f1}. The result is

$$t_{f1} = \tau_q \ln \frac{I_{B0} - I_{CS}/h_{FE}}{I_{B0}}, \quad (11.18b)$$

where

$$\tau_q = \frac{h_{FE}}{\omega_T} \quad (11.18c)$$

From (11.17) and (11.18a) we may obtain

$$v_{CE}(t_{f1}) = V_{CEX}. \quad (11.19)$$

During t_{f2}, $V_{CEX} \leq v_{CE}(t) \leq 0.9(V_{CE0} - V_{CES}) + V_{CES}$, and the Kirchhoff equation for the output circuit is

$$\frac{V_{CC} - v_{CE}}{R_C} = C_L \frac{dv_{CE}}{dt} + \frac{v_{CE}}{R_L}. \quad (11.20)$$

The solution to (11.20) yields

$$t_{f2} = \tau_{f2} \ln \left[\frac{I_{CS} - V_{CEX}/R_{EQ}}{0.1 I_{CS}} \right], \quad (11.21)$$

where

$$\tau_{f2} = R_{EQ} C_L. \quad (11.22)$$

If t_{f1} is calculated using (11.18b), $V_{CEX} \simeq V_{CES}$.

From the foregoing results, the time constants that are associated with the various switching intervals of the inverter are

$$\tau_d = R_B(C_{BE} + C_{CB}) \quad (11.9b)$$

$$\tau_r = \frac{h_{FE}}{\omega_T} + \frac{R_C R_L}{R_C + R_L}(h_{FE} C_{CB} + C_L) \quad (11.12b)$$

$$\tau_s \simeq \frac{1}{\omega_{\alpha I}(1 - \alpha_I)} \quad (11.15)$$

$$\tau_{f1} = \tau_r \quad (11.16)$$

$$\tau_q = \frac{h_{FE}}{\omega_T} \quad (11.18c)$$

and

$$\tau_{f2} = \frac{R_C R_L}{R_C + R_L} C_L. \quad (11.22)$$

For a representative micropower case where $R_B = 3 \times 10^4 \, \Omega$, $C_{BE} \simeq 5 \, \text{pF}$, $C_{CB} \simeq 1 \, \text{pF}$, $C_L \simeq 50 \, \text{pF}$, $h_{FE} = 50$, $\omega_T = 0.66 \times 10^9$, $R_C = 3 \times 10^4 \, \Omega$,

Bipolar Transistor Digital Circuits

$R_L = 3 \times 10^4 \, \Omega$, $\omega_{\alpha I} = 0.2 \times 10^8$, and $\alpha_I \simeq 0.1$, $V_{BET} = 0.6$ V, $I_{BS} = 0.01$ ma, $I_{B0} = -0.01$ ma, $V_{CC} = 3.0$ V, the values of these time constants are

$$\tau_d = 1.8 \times 10^{-7} \text{ sec}$$
$$\tau_r = 0.75 \times 10^{-7} + 150 \times 10^{-7} \text{ sec}$$
$$\tau_s = 0.55 \times 10^{-7} \text{ sec} \quad (11.23a)$$
$$\tau_{f1} = 0.75 \times 10^{-7} + 150 \times 10^{-7} \text{ sec}$$
$$\tau_g = 0.75 \times 10^{-7} \text{ sec}$$

and

$$\tau_{f2} = 75 \times 10^{-7} \text{ sec}$$

for which

$$t_d = 1.98 \times 10^{-7} \text{ sec}$$
$$t_r = 29.8 \times 10^{-7} \text{ sec}$$
$$t_s = 0.28 \times 10^{-7} \text{ sec} \quad (11.23b)$$
$$t_{f1} = 0.14 \times 10^{-7} \text{ sec}$$
$$t_{f2} = 172 \times 10^{-7} \text{ sec.}$$

These results suggest that the average switching time of a micropower inverter

$$t_{sw} = \tfrac{1}{2}(t_d + t_r + t_s + t_f) \quad (11.24)$$

is dominated by circuit time constants and is relatively independent of the intrinsic capability of the transistor used in the circuit. In fact,

$$t_{sw} \simeq \tfrac{1}{2}(t_r + t_{f2}) \quad (11.25)$$

with

$$\tau_r \simeq \frac{R_C R_L}{R_C + R_L}(h_{FE}C_{BC} + C_L) \quad (11.26a)$$

and

$$V_{CEX} \simeq V_{CES} \quad (11.26b)$$

is often an accurate approximation for (11.24) in the micropower range. Two salient features of micropower inverter operation implied by (11.25) are the negligible contributions of transistor storage time t_s and active region fall time t_{f1} to the average switching time. Frequently, only rise time t_r and cutoff region fall time t_{f2} are significant.

Combining (11.7b) and (11.25) gives

$$P^*_{DC} \simeq K_P V_{CC}^2 \frac{C_L}{t_{sw}} \quad (11.27)$$

for

$$K_P = \frac{1}{4}\frac{R_L}{R_C + R_L}\left[\left(1 + h_{FE}\frac{C_{CB}}{C_L}\right)\ln\left(\frac{h_{FE}I_{BS}}{h_{FE}I_{B\bar{S}} - 0.9I_{CS}}\right) + 2.3\right], \quad (11.28)$$

assuming $C_L \gg C_{CB}$ and $[(R_C + R_L)/R_C R_L]V_{CES} \ll I_{CS}$. From (11.27) it is evident that the average power consumption of the micropower inverter, P_{DC}^*, is inversely proportional to its switching speed t_{sw}.

If the circuit elements external to the transistor are stripped from the inverter in Figure 11.1, the switching speed of the transistor itself remains. For $1/R_L = 1/R_C = C_L = 0$ and a current source drive in the base circuit

$$t_d = \frac{C_{BE} + C_{CB}}{I_{BS}}(V_{BET} - V_{B0}) \quad (11.29a)$$

$$t_r = \frac{h_{FE}}{\omega_T}\ln\left(\frac{h_{FE}I_{BS}}{h_{FE}I_{BS} - 0.9I_{CS}}\right) \quad (11.29b)$$

$$t_s = \tau_s \ln\frac{I_{BS} - I_{B0}}{I_{CS}/h_{FE} - I_{B0}} \quad (11.29c)$$

and

$$t_f = t_{f1} = \frac{h_{FE}}{\omega_T}\ln\left(\frac{I_{CS} - h_{FE}I_{B0}}{0.1I_{CS} - h_{FE}I_{B0}}\right), \quad (11.29d)$$

where

$$\omega_T = \frac{g_m}{C_{BE} + C_{CB}} = \frac{\gamma I_C}{C_{BE} + C_{CB}}. \quad (11.30)$$

For $I_{CS}/I_{BS} = |I_{CS}/I_{B0}| = 10$, Figure 11.5 illustrates the switching behavior of a typical micropower transistor. Essentially, the calculations are based on (11.29) and (11.30), assuming average capacitances C_{BE} and C_{CB} within a given switching interval. These capacitances, which consist principally of emitter and collector junction capacitance, respectively, dominate the switching speed and gain-bandwidth product of a transistor in the micropower range. Since device voltage swings are essentially unchanged in the micropower range, the charge demands of the junction capacitances for switching operation are fixed. The small device currents, however, require a longer time interval to charge the junction capacitances. This results in the diminution of device switching speed in nearly direct proportion to collector current reduction as indicated in Figure 11.5 [8]. The storage times indicated by the t_s curve in Figure 11.5 consist essentially of the 0 to 10% portion of the fall time and are calculated from a modified form of (11.29d). The usual minority carrier storage time, as described by (11.29c), is insignificant as a

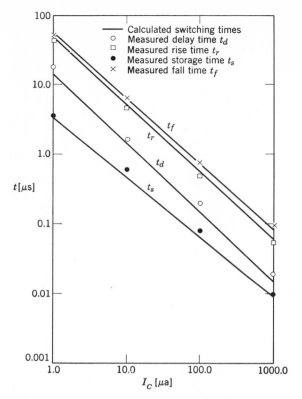

Figure 11.5 Micropower transistor calculated and measured turn-on delay time t_d, rise time t_r, storage time t_s, and fall time t_f, versus saturated collector current I_C for $I_C/I_{BS} = |I_C/I_{BO}| = 10$, $V_{CC} = 3.0$ V, $T = 25°$C.

result of the negligible base region charge storage accompanying the reduced device currents of the micropower range.

On the basis of (11.27) and Figure 11.5, it is quite clear that a power-speed trade-off can be made in the design of digital circuits. In most instances no design constraint of a digital logic circuit influences its power consumption as markedly as the speed of operation, although virtually all design constraints do exert some influence on the power level. Consequently, it is evident that in micropower applications the operating speeds of digital circuits should not exceed system requirements.

The transistors selected for use in micropower digital circuits should be extremely small geometry devices capable of nanosecond switching speeds at relatively large operating currents. By reducing the operating currents of such nanosecond switching transistors, microsecond switching can be achieved at drastically reduced power levels, as Figure 11.5 indicates. At the same time, certain minimal dc requirements must be satisfied. Specifically, the most basic micropower transistor requirements are [8] (a) junction reverse currents that at maximum device temperatures are small compared with minimum operating currents, (b) useable device current gains at minimum operating currents and temperatures, and (c) minimum junction capacitances in order to enhance switching speed.

Finally, the type of digital circuit in which a micropower transistor is imbedded and the nature of the circuit environment, especially with regard to stray wiring capacitance and noise, strongly influence power consumption. For example, the effect of C_L is readily apparent in (11.27). On this basis alone the conditions of a microcircuit environment in which isolation capacitances, packaging capacitances, wiring capacitances, and other forms of stray capacitance are minimized can be highly favorable to micropower operation of digital circuits.

11.2 AN ITERATIVE CHAIN OF INVERTERS

In its skeletal form the logic circuitry of a digital system often can be represented by a simple chain of inverters, as illustrated in Figure 11.6. The Kirchhoff loop equations that describe the quiescent state of an iterative pair of these inverters are

$$V_{CES} = I_{B0}R_B + V_{BE0}, \tag{11.31a}$$

$$V_{CC} = (I_{C0} + I_{BS})R_C + V_{CE0}, \tag{11.31b}$$

$$V_{CE0} = I_{BS}R_B + V_{BES}, \tag{11.31c}$$

$$V_{CC} = (I_{CS} + I_{B0})R_C + V_{CES}, \tag{11.31d}$$

where the subscript 0 applies to a cutoff transistor and S to a saturated device.

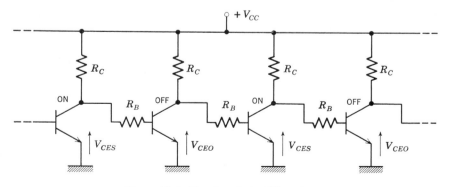

Figure 11.6 Iterative chain of inverters.

In order to maintain alternately cutoff and saturated inverters in the chain, the requirements can be written as

$$V_{BET} \geq V_{BE0} \tag{11.32}$$

for the cutoff inverter and

$$I_{BS} \geq \frac{I_{CS}}{h_{FE}} \tag{11.33}$$

for the saturated inverter where the threshold voltage V_{BET} is slightly less than V_{B2} in Figure 11.2a.

Combining (11.31a) and (11.32) gives

$$V_{BET} \geq V_{CES} - I_{B0}R_B \simeq V_{CES} \tag{11.34}$$

for $V_{CES} \gg I_{B0}R_B$ as the requirement for a stable cutoff state. From (11.31b), (11.31c), and (11.33),

$$R_B \leq \frac{V_{CC} - V_{BES} - I_{C0}R_C}{I_{CS}/h_{FE}} - R_C \tag{11.35}$$

is the requirement for a stable saturated state. It can be assumed that I_{CS} is determined by switching speed constraints and that (11.31d) then defines R_C, since $I_{B0} \ll I_{CS}$ and $V_{CES} \ll V_{CC}$.

From (11.34),

$$\Delta V_0 \simeq V_{BET} - V_{CES} \tag{11.36}$$

can be considered the cutoff stability margin of an inverter. It is the difference between the maximum allowable input voltage to a cutoff inverter, V_{BET}, and the actual input voltage, V_{CES}.

From (11.31c),

$$\Delta V_S = R_B\left(I_{BS} - \frac{I_{CS}}{h_{FE}}\right) \tag{11.37}$$

can be considered the saturated stability margin. It is the difference between the actual input voltage of a saturated inverter, $V_{CE0} = I_{BS}R_B + V_{BES}$, and the minimum allowable input voltage to a saturated inverter, $V_{CE0} = (I_{CS}/h_{FE})R_B + V_{BES}$. If we select $I_{BS} = \theta I_{CS}/h_{FE}$ with $\theta > 1$ in accordance with (11.33), then (11.35) gives

$$R_B = \frac{V_{CC} - V_{BES} - I_{C0}R_C}{\theta I_{CS}/h_{FE}} - R_C \tag{11.38}$$

and

$$\Delta V_S = R_B \frac{I_{CS}}{h_{FE}}(\theta - 1) \tag{11.39}$$

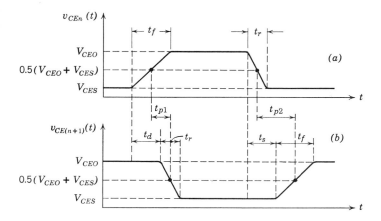

Figure 11.7 Inverter output waveforms for successive stages.

from (11.37). For the inverter chain in Figure 11.6, it is clear from (11.36) that ΔV_0 depends on transistor parameters that are practically nonadjustable. This is a particular problem in the micropower range at high temperatures. For example, in a typical case [8] at $T = 125°C$, $V_{BET} \simeq 0.43$ V and $V_{CES} \leq 0.14$ V for $(V_{BET} - V_{CES}) = 0.29$ V at $I_C = 10 I_B = 1.0$ mA, whereas $V_{BET} = 0.24$ V and $V_{CES} = 0.10$ V for $(V_{BET} - V_{CES}) = 0.14$ V at $I_C = 10 I_B = 10$ μA. On the other hand, ΔV_S is easily controlled by means of the resistance R_B or the transistor current gain h_{FE}.

The dynamic behavior of the chain of inverters in Figure 11.6 is illustrated by the waveforms in Figure 11.7. The nature of the driving waveform and the load of a given inverter are more complex than for the single inverter of the previous section. The most important measure of the operating speed of the chain of inverters is the average propagation delay time

$$t_{pd} = \tfrac{1}{2}(t_{p1} + t_{p2}). \tag{11.40}$$

From Figure 11.7,

$$t_{p1} = t_d - \frac{t_f}{2} + \frac{t_r}{2} \tag{11.41a}$$

and

$$t_{p2} = t_s - \frac{t_r}{2} + \frac{t_f}{2} \tag{11.41b}$$

so that

$$t_{pd} = \tfrac{1}{2}(t_d + t_s). \tag{11.42}$$

In the micropower range

$$t_d \leq t_f \quad \text{and} \quad t_s \leq t_r \tag{11.43}$$

in virtually all cases. Consequently, in such instances

$$t_{pd} \leq \tfrac{1}{2}(t_r + t_f). \tag{11.44}$$

With the aid of a programmed computer, a useful approximation for t_{pd} can be calculated conveniently, essentially by reducing the problem to a piecewise linear analysis equivalent to the single inverter switching analysis of the previous section. Using the single inverter circuit in Figure 11.1 as a convenient reference, t_r in Figure 11.7 for the chain of inverters must be divided into two intervals, t_{r1} during which the "driven" inverter is active and t_{r2} during which it is cutoff. During t_{r1}, the driving inverter collector voltage swing is $V_{CE0} = I_{BS}R_B + V_{BES} \geq v_{CE}(t) \geq (I_{CS}/h_{FE})R_B + V_{BES}$, the driving inverter base current is $I_{BS} \simeq (V_{CC} - V_{BES})/(R_B + R_C)$, the driven capacitance C_L consists of wiring capacitance C_W from driving inverter output to driven inverter input, and R_L consists of R_B in series with a battery voltage V_{BES}. During t_{r2}, the driving inverter collector voltage swing is $(I_{CS}/h_{FE})R_B + V_{BES} \geq v_{CE}(t) \geq 0.1(V_{CE0} - V_{CES}) + V_{CES}$, the base drive is unchanged, the capacitance C_L consists approximately of $(C_W + C_{BE} + C_{CB})$, and $1/R_L \simeq 0$.

The fall time t_f for a single inverter must be divided into two intervals. During t_{f1} the inverter is active, during t_{f2} it is cutoff. For the chain of inverters, these two intervals of the driving inverter must be meshed with the two intervals of the driven inverter, which is initially cutoff and then becomes active during the driving inverter fall time. Consequently two possible three-interval sequences may arise during t_f for the chain of inverters.

FALL TIME SEQUENCE NO. 1

Driving Inverter:	active	active	cutoff
Driven Inverter:	active	cutoff	cutoff

FALL TIME SEQUENCE NO. 2

Driving Inverter:	active	cutoff	cutoff
Driven Inverter:	active	active	cutoff

These three interval sequences may degenerate into two or one interval fall times. Regardless of the number of intervals involved, however, simple linear models for the circuit during each interval can be defined, analyzed, and pieced together to predict overall dynamic behavior. Although this procedure is somewhat cumbersome, by programmed computation it permits useful predictions of the operating speed of rather complex arrangements of micropower logic circuits.

11.3 RESISTOR TRANSISTOR LOGIC (RTL)

The most common logic gate is an NPN bipolar transistor circuit. Figure 11.8 illustrates the schematic diagrams of the basic circuit configurations of six well-established forms of saturating logic gates: (a) resistor transistor logic (RTL), (b) transistor resistor logic (TRL), (c) diode transistor NOR logic (DTL NOR), (d) diode transistor NAND logic with resistor level shifting (DTL NAND R), (e) diode transistor logic with diode level shifting (DTL NAND D), and (f) transistor transistor logic (TTL). As the operating power levels of these basic gates are reduced to the micropower range, their usual performance capabilities change significantly. The purpose of this section is to begin (a) to discuss these changes, (b) to assess the relative micropower performance of the gates, and (c) to establish a firm basis for comparison of the micropower potential of more advanced types of gates.

Because of its longstanding utility [21–23], compatibility with discrete component, hybrid thin film, and silicon monolithic fabrication technologies as well as its adaptability to micropower operation [8,24,25], the RTL circuit in Figure 11.8a provides an excellent vehicle for demonstration of the optimum design of a micropower basic gate.

Worst Case Analysis

In order to ensure reliable operation of a basic gate, it must be designed to withstand specified manufacturing and environmental tolerances in the values of the circuit elements, variations in supply voltages, and spurious noise voltages. A specific degree of immunity to these disturbances must be assured for the most stringent conditions of input (fan-in) and output (fan-out) loading under which the gate is expected to function. A design technique that is quite useful in accomplishing this objective is "worst case design" [1,2,6]. Worst case design is based on the particular combination of parameter values that will result in an extreme value of a desired variable in the direction that will make the circuit least likely to operate correctly. For example, the combination of parameter values for V_{CC}, R_C, R_B, and V_{BES} resulting in minimum saturated base current, I_{BS}, as described by (11.33), would be used in a worst case design.

A schematic diagram that illustrates worst case conditions for a RTL gate is shown in Figure 11.9. The Kirchhoff loop equations that describe the circuit are

$$\bar{V}_{CES} = \underline{I}_{B0}R_1 + \bar{V}_{BE0}, \tag{11.45a}$$

$$\underline{V}_{CC} = [M\bar{I}_{C0} + (N-1)\bar{I}_{BS} + \underline{I}_{BS}]\bar{R}_C + \underline{I}_{BS}\bar{R}_1 + \bar{V}_{BES}, \tag{11.45b}$$

$$0 = \underline{I}_{BS}\bar{R}_1 + \bar{V}_{BES} - \bar{I}_{BS}\underline{R}_1 - \underline{V}_{BES}, \tag{11.45c}$$

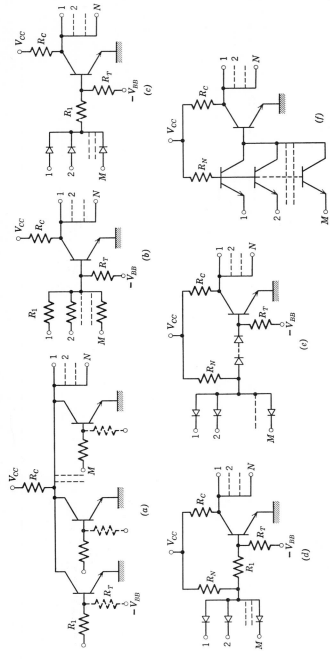

Figure 11.8 Schematic diagrams of basic logic gates: (*a*) resistor transistor logic (RTL); (*b*) transistor resistor logic (TRL); (*c*) diode transistor NOR logic (DTL NOR); (*d*) diode transistor NAND logic with resistor level shifting (DTL NAND R); (*e*) diode transistor logic with diode level shifting (DTL NAND D); (*f*) transistor transistor logic (TTL).

Resistor Transistor Logic (RTL)

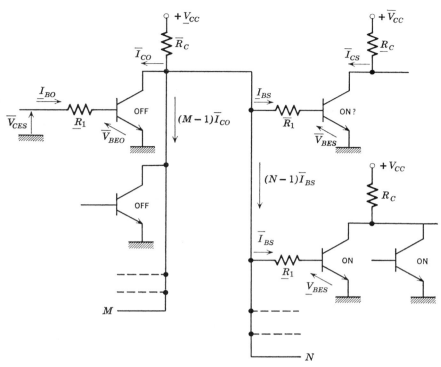

Figure 11.9 Schematic diagram of worst case conditions for a RTL gate.

and
$$\bar{V}_{CC} = \bar{I}_{CS}\underline{R}_C + \underline{V}_{CES}, \qquad (11.45d)$$

where a super bar [e.g., $\bar{V}_{CES}$, $\bar{V}_{CC} = V_{CC}(1 + \delta_V)$ and $\bar{R}_C = R_C(1 + \delta_R)$] denotes the maximum value of a parameter, a sub bar [$\underline{V}_{CES}$, $\underline{V}_{CC} = V_{CC}(1 - \delta_V)$ and $\underline{R}_C = R_C(1 - \delta_R)$] denotes the minimum value, and the absence of both (V_{CC} and R_C) denotes a nominal value. The decimal tolerance of a quantity is indicated by δ (e.g., $\delta_V = 0.05$ indicates a $+5\%$ tolerance in V_{CC}). (Note that I_{C0} is the value of I_C of a cutoff transistor and not I_{CB0}.)

In (11.45a) it is assumed that the direction of I_{B0} is into the base. At high temperatures for transistors with $(1 - \alpha_N)I_{ES} \ll (1 - \alpha_I)I_{CS}$, collector junction reverse current may dominate I_{B0} and therefore reverse its sign, as (11.16) suggests. For $I_{B0} > 0$, however, it is clear that the combination of minimum values $\underline{I}_{B0}$ and $\underline{R}_1$ and the maximum value $\bar{V}_{CES}$ result in the extreme worst case value $\bar{V}_{BE0}$. As (11.5) suggests, $\bar{V}_{CES}$ in (11.45a) normally corresponds to the maximum operating temperature, since $\gamma = kT/q$, and the maximum value of I_C/I_B. The worst case condition that must be met for a stable cutoff gate is

$$\underline{V_{BET}} \geq \bar{V}_{BE0}. \qquad (11.46)$$

The threshold voltage V_{BET} is not uniquely defined as a transistor parameter since the $I_C(V_{BE})$ characteristic is exponential and "the" voltage at which I_C just begins to flow is problematical. If, however, we define I_{CT} as the collector current corresponding to V_{BET}, (11.1d) gives

$$V_{BET} = \frac{nkT}{q} \ln \frac{I_{CT}}{\alpha_N I_{EBS}} \qquad (11.47)$$

and from (11.45b)

$$\underline{V_{BET}} = \frac{nkT}{q} \ln \left[\frac{\underline{V_{CC}} - \bar{V}_{BES} - \underline{I_{BS}}(\bar{R}_1 + \bar{R}_C) - (N-1)\bar{I}_{BS}\bar{R}_C}{M\bar{R}_C \alpha_N I_{EBS}} \right]. \qquad (11.48)$$

On the basis of (11.48), $\underline{V_{BET}}$ is defined as the base-to-emitter voltage of the cutoff gate at which it just maintains all its output or fan-out loads in a saturated condition [8]. Clearly, $\underline{V_{BET}}$ corresponds to maximum fan-in (M) for the cutoff gate. In addition, $\underline{V_{BET}}$ should be taken at the upper operating temperature limit of the circuit.

The tolerances reflected in (11.45b) and (11.45c) are such that minimum base current flows in one "starved" gate that must maintain a base current of at least $\bar{I}_{CS}/h_{FE}$ to remain saturated. It is evident that $\underline{V_{CC}}$ and $\bar{R}_C$ reduce the total current available at the fan-out node and that $(N-1)$ fan-out gates each with $\underline{R}_1$ and $\underline{V_{BES}}$ present a worst case for current hogging from the starved gate. From (11.1b) in the saturation region

$$\underline{V_{BES}} = \frac{nkT}{q} \ln \left[\frac{\underline{I_{BS}}}{(1-\alpha_N)I_{EBS} + (1-\alpha_I)I_{CBS}\epsilon^{-V_{CES}/\gamma}} \right] \qquad (11.49)$$

indicates that $\underline{V_{BES}}$ for the current-hogging transistors corresponds to the worst case condition where V_{CES} of these transistors is clamped by the saturated transistors of the remaining inputs to the current-hogging gates, as illustrated in Figure 11.9. Since all saturated gates are not necessarily at the same temperature, a reasonable temperature differential should be assumed between $\bar{V}_{BES}$ and $\underline{V_{BES}}$ in addition to the other factors contributing to the difference between the two values.

The worst case condition that must be met for a stable saturated gate is

$$\underline{I_{BS}} \geq \frac{\bar{I}_{CS}}{h_{FE}}. \qquad (11.50)$$

Normally this condition is most difficult to achieve at the lower operating temperature limit of the circuit, because $h_{FE}(T)$ is essentially a monotonically increasing function. In the micropower range, however, the value of $M\bar{I}_C$, in (11.45b) can become significant at high temperatures so that the base drive of the starved gate is smallest at the upper operating temperature limit of the circuit [8]. Consequently the worst case design must consider both temperature limits in the micropower range.

From (11.45d), the required value of $\bar{I}_{CS}$ for use in (11.50) can be determined. In a micropower design the nominal value I_{CS} is determined principally by the required operating speed of the circuit, which consequently sets the operating power level as discussed previously. With I_{CS} determined, (11.45d), in nominal form, yields R_C.

For $\underline{I}_{BS} = \theta I_{CS}/h_{FE}$, $\theta > 1$ in accordance with (11.50), we can proceed by selecting M and N and solving (11.45b and 11.45c) for

$$R_1^2 + K_1 R_1 + K_2 = 0, \tag{11.51}$$

where

$$K_1 = \frac{[(M\bar{I}_{C0} + \underline{I}_{BS})\bar{R}_C + \bar{V}_{BES} - \underline{V}_{CC}](1 - \delta_R) + (N-1)\underline{I}_{BS}\bar{R}_C(1 + \delta_R)}{\underline{I}_{BS}(1 + \delta_R)(1 - \delta_R)}$$

$$\tag{11.52a}$$

and

$$K_2 = \frac{(N-1)\bar{R}_C(\bar{V}_{BES} - \underline{V}_{BES})}{\underline{I}_{BS}(1 + \delta_R)(1 - \delta_R)}, \tag{11.52b}$$

from which the nominal range of values of R_1, allowable for the worst case design, can be obtained. The value

$$R_1 = \frac{V_{CC} - (MI_{C0} - NI_{BS})R_C - V_{BES}}{I_{BS}} \tag{11.53}$$

corresponds to the nominal design for which $K_1 = -R_1$ and $K_2 = 0$.

By solving (11.45b and 11.45c) for

$$N = \left[\frac{\underline{V}_{CC} - (M\bar{I}_{C0} + \underline{I}_{BS})\bar{R}_C - \underline{I}_{BS}R_1(1 + \delta_R) - \bar{V}_{BES}}{\bar{V}_{BES} - \underline{V}_{BES} + \underline{I}_{BS}R_1(1 + \delta_R)}\right]\frac{R_1(1 - \delta_R)}{\bar{R}_C} + 1$$

$$\tag{11.54}$$

and taking the derivative $\partial N/\partial R_1$, we get

$$R_1^2 + 2\frac{(\bar{V}_{BES} - \underline{V}_{BES})}{\underline{I}_{BS}(1 + \delta_R)}R_1$$

$$- \frac{[\underline{V}_{CC} - (M\bar{I}_{C0} + \underline{I}_{BS})\bar{R}_C - \bar{V}_{BES}](\bar{V}_{BES} - \underline{V}_{BES})}{[\underline{I}_{BS}(1 + \delta_R)]^2} = 0, \tag{11.55}$$

which can be solved for the optimum value of R_1 for which N is maximum.

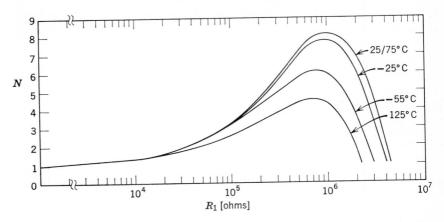

Figure 11.10 Fan-out versus base padding resistance for a worst case RTL circuit.

Figure 11.10 illustrates the dependence of fan-out on base padding resistance R_1 for a typical set of design constraints for the micropower range. For a given fan-out the lower limit on R_1 is set by current hogging of transistors with low-input thresholds, whereas the upper limit is set by R_1 itself reducing base drive to the minimum allowable value. The basis for choosing R_1 includes both maximum fan-out and operating speed. If a very large fan-out and consequently large values of R_1 are necessary, a "speed up" capacitor can be used in parallel with R_1 (see Figure 11.15). The switching speed of the resulting resistor capacitor transistor logic (RCTL) circuit is improved, but its susceptibility to ac noise increases.

The worst case dc stability margin for a cutoff RTL gate is

$$\Delta V_0 = V_{BET} - \bar{V}_{BE0}, \tag{11.56}$$

where (11.45a) gives $\bar{V}_{BE0}$ and (11.48) gives V_{BET}. In essence, this stability margin is taken as the difference between the maximum possible input voltage at which the cutoff gate can maintain its output or fan-out loads in a saturated condition and the worst case dc input voltage to the cutoff gate.

For a saturated gate the worst case stability margin is

$$\Delta V_S = \bar{R}_1 \left(I_{BS} - \frac{\bar{I}_{CS}}{h_{FE}} \right), \tag{11.57}$$

where the allowable range for R_1 is given by (11.51), $\bar{I}_{CS}$ by (11.45d), and $I_{BS} = \theta \bar{I}_{CS}/h_{FE}$, $\theta > 1$ is selected to provide reasonable overdrive for rapid turn-on and to keep the starved gate deep in saturation. The saturated stability margin is the difference between the worst case dc input voltage to a saturated gate and the minimum possible input voltage at which the gate remains saturated.

The average power drain associated with a worst case RTL circuit is [8]

$$P_{DC}^* = pP_S + (1 - p)P_{C0}, \qquad (11.58)$$

where, from Figure 11.9, the power associated with a saturated gate is

$$P_S \simeq \bar{V}_{CC} I_{CS} \qquad (11.59)$$

and with a cutoff gate,

$$P_{C0} \simeq \underline{V_{CC}}[M\bar{I}_{C0} + (N-1)\bar{I}_{BS} + I_{BS}]. \qquad (11.60)$$

The probability $p(M) = p = 1 - p^M$ of a gate being saturated depends on the fan-in of the gate. Representative values of $p(M)$ are listed:

$p(1) = 0.500 \qquad p(5) = 0.755$
$p(2) = 0.618 \qquad p(6) = 0.778$
$p(3) = 0.682 \qquad p(7) = 0.797$
$p(4) = 0.724 \qquad p(8) = 0.812$

From these values the degree of error in the usual approximation for power drain,

$$P_{DC}^* = \tfrac{1}{2}(P_S + P_{C0}), \qquad p = 0.500, \qquad (11.61)$$

is evident.

Using a piecewise linear analysis similar to that discussed for the chain of inverters in the previous section and programmed computation, worst case dynamic behavior of the RTL circuit can be described in detail. Since the procedure for such an analysis has been established and the details are of little interest, it will not be pursued further. A suitable worst case RTL circuit for analysis of dynamic performance is one in which (a) the driving gate is switched by only one of its M fan-in transistors while the other $(M - 1)$ remain cutoff, and (b) the driving gate fan-out is N driven gates that are switched by the driving gate.

At this point it is apparent that there is a significant degree of interaction among fan-in, fan-out, cutoff stability margin, saturated stability margin, tolerance immunity, operating temperature range, propogation delay time, and average power consumption for a RTL gate. The interdependence of these quantities is quite complex and thus not easily described in simple analytical terms. With each design constraint defined, we can best proceed toward an optimum micropower circuit design by selecting a likely value for I_{CS}, since specification of this current and the collector supply voltage V_{CC} largely define the associated range of operating points of the transistor and hence its pertinent performance characteristics. Computer computation of a large number of trial designs based on the analytical results previously discussed then permits the identification of a suitable optimum design.

Performance Trade-Offs

A graphic illustration of some of the major trade-offs involved in the optimum design of a RTL gate is displayed in Figure 11.11. The design constraints on which Figure 11.11 is based are rather severe. In particular, the operating temperature range, $-55°C \leq T \leq 125°C$, at $I_{CS} = 10$ μA

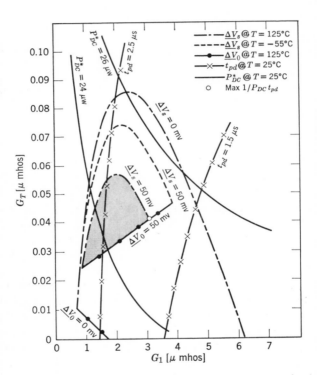

Figure 11.11 Base turn-off conductance ($G_T = 1/R_T$) versus series base conductance ($G_1 = 1/R_1$) for a RTL gate. Saturated transistor collector current $\bar{I}_{CS} = 10.0$ μa. Operating temperature range, $-55°C \leq T \leq 125°C$; fan-in (M) = fan-out (N) = 3; 5% resistor (R_1, R_T, and R_C) tolerance ($\delta_R = \pm 0.05$); supply voltage $V_{CC} = -V_{BB} = 3.0$ V with 5% tolerance ($\delta_V = \pm 0.05$). Loci of constant average power drain P_{DC}^*, constant average propogation delay time t_{pd}, constant cutoff stability margin $\underline{\Delta V_0}$ and saturated stability margin $\underline{\Delta V_S}$ are indicated. Shaded area represents design range for stability margin $\underline{\Delta V} \geq$ 50 mV. Optimum design for maximum $1/P_{DC}^* t_{pd}$ is indicated.

combined with the required dc stability margins, $\underline{\Delta V_0} \geq 50$ mV and $\underline{\Delta V_S} \geq 50$ mV, requires the use of a negative supply voltage, $-V_{BB}$, as illustrated in Figure 11.8a. The additional circuit complexity this entails is undesirable especially because of the large values of turn-off resistance $R_T = 1/G_T$ that may be required in integrated circuits. Nevertheless the curves of Figure 11.11 are useful in illustrating trade-offs in the design of micropower transistor logic [8].

Based on equations similar to those already discussed in this section, Figure 11.11 displays loci of constant worst case stability margin $\underline{\Delta V}$, constant average power drain P_{DC}^*, and constant average propagation delay time t_{pd}. The design constraints are summarized in the figure legend. In addition, the designs assume a ± 50 mV manufacturing tolerance and a 20°C temperature differential for base-emitter diodes of transistors in a current-hogging situation. By plotting conductances (that is, G_T versus G_1), the $\underline{\Delta V_0}$ loci become straight lines. To insure $\underline{\Delta V_0} \geq 50$ mV, G_T for a cutoff gate must take on a value above the locus $\underline{\Delta V_0} \gtrless 50$ mV at 125°C. Also, to insure $\underline{\Delta V_S} \geq 50$ mV, G_T for a saturated gate must assume a value beneath the locus for $\underline{\Delta V_S} = 50$ mV at 125°C. Consequently, for $\underline{\Delta V_0} \geq 50$ mV and $\underline{\Delta V_S} \geq 50$ mV the allowable design range for G_T and G_1 is the shaded area. Note that since the locus for $\underline{\Delta V_S} = 50$ mV at 125°C lies beneath the locus for $\underline{\Delta V_S} = 50$ mV at $T = -55$°C, the $\underline{\Delta V} \geq 50$ mV design range is exclusively high temperature limited, which is largely a micropower anomaly.

An optimum design for $\underline{\Delta V} \geq 50$ mV in terms of a maximum value of the reciprocal power drain-propagation delay time product $1/P_{DC}^* t_{pd}$ is illustrated by the encircled point in the extreme right corner of the shaded design area. Its rather minimal worst case design constraint of $\underline{\Delta V} \geq 50$ mV may be adequate in many micropower applications, since overall internal noise generation is diminished at micropower levels. $L(di/dt)$ ground plane noise is reduced by smaller Δi and larger Δt but $RC(dv/dt)$ crosstalk noise is not reduced since Δv does not change and R increases in proportion to the Δt increase as operating current decreases. External noise is generally unchanged for micropower circuits, and improved shielding is necessary. Fortunately this is assisted by the increased packaging densities permissible at micropower levels. The maximum operating temperature of a RTL gate must be reduced to about 100°C to achieve the desirable objective of $\underline{\Delta V} \gtrless 50$ mV at $I_{CS} = 10$ μA without the use of a negative supply voltage. By further limiting the temperature range to -25°C $\leq T \leq 75$°C, an expanded logic capability, $M = 9$ and $M = 4$, is possible.

The speed-power performance of a series of RTL gates of optimum design, as described in connection with Figure 11.11, is illustrated in Figure 11.12. Based on essentially similar optimum design techniques [8], the speed-power performance of the remaining basic gates of Figure 11.8 is illustrated in

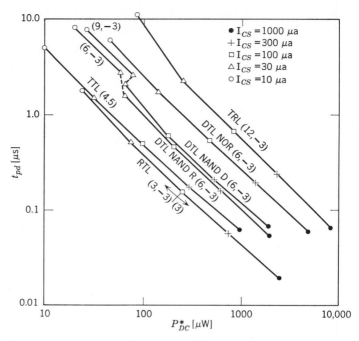

Figure 11.12 Average propagation delay time t_{pd} versus average power drain P_{DC}^* for various micropower basic gates. Supply voltages are indicated in parenthesis as (V_{CC}, V_{BB}) or (V_{CC}). Operating temperature range, $-55°C \leq T \leq 125°C$; fan-in (M) = fan-out $(N) = 3$; $\pm 5\%$ resistor and supply voltage tolerances; stability margin $\underline{\Delta V} \geq 50$ mV except for TTL ΔV_0 for $I_{CS} < 100$ μa; DTL NAND R uses $1/R_C = 0$ for $\overline{I_{CS}} \geq 30$ μa only; DTL NAND D uses three-level shifting junctions for $I_{CS} \leq 30$ μa only; all designs assume worst case conditions and are optimized for maximum $1/P_{DC}^* t_{pd}$.

Figure 11.12 for comparative purposes. The nearly unity slopes of these curves indicate that the key speed-power products remain essentially constant throughout an extended micropower range. For $I_{CS} < 10$ μA, usable circuit designs become quite difficult to achieve subject to the constraints summarized in the legend of Figure 11.12.

The RTL gate offers the best speed-power performance. However, it loses the key advantage of simplicity for $I_{CS} < 100$ μa where a negative supply voltage becomes necessary to provide adequate $\underline{\Delta V}$. In applications where the maximum operating temperature is sufficiently low to permit adequate $\underline{\Delta V_0}$ for $1/R_T = 0$ and $V_{BB} = 0$, the RTL gate offers attractive micropower performance. (The remaining curves of Figure 11.12 are discussed in the latter portions of this section.) Table 11.1 describes the design details of a group of micropower basic gate circuits.

Resistor Transistor Logic (RTL)

Table 11.1 Micropower Logic Circuit Designs
$(\bar{I}_{CS} = 10\ \mu\text{a},\ M = N = 3,\ \Delta V \geq 50\ \text{mV})$

	T(min) (°C)	T(max) (°C)	V_{CC} (V)	V_{BB} (V)	R_C (MΩ)	R_1 (MΩ)	R_T (MΩ)	R_N (MΩ)	ΣR (MΩ)	P_{DC}^* (μw)	t_{pd} (μs)	$P_{DC}^* t_{pd}$ (μw-μS)
RTL	−55	+125	3	3	0.32	0.32	24	—	73	25	1.8	45
	−25	+75	3	—	0.32	0.15	—	—	0.77	25	1.2	30
TRL	−55	+125	12	3	1.4	3.0	70	—	80	87	11.2	970
	−25	+75	12	—	1.5	1.6	—	—	6.3	85	6.8	570
DTL NOR	−55	+125	6	3	0.66	0.54	14	—	15	47	5.8	270
	−25	+75	6	3	0.68	0.45	9.1	—	10	49	4.4	210
DTL NAND-R	−55	+125	6	3	6.5	0.86	4.9	1.9	14	19	8.8	170
	−25	+75	6	3	—	0.59	3.6	1.6	5.8	21	4.5	94
DTL NAND-D	−55	+125	9	3	1.7	3	—	3.8	2.8	8.3	29	11.5	340
	−25	+75	6	3	—	2	14	1.8	16	18	4.7	85
TTL	−55	+125	4.5	—	1.3	—	—	3.2	4.5	10	5.0	50[a]
	−25	+75	3	—	1.3	—	—	1.4	2.7	6.7	3.2	21

[a] $\Delta V_0 \geq 20$ mV.

Storage Circuits

The basic flip-flop or storage circuit is formed in the RTL system, as in most others, by cross-coupling two basic gates as illustrated in Figure 11.13. The truth table for the *R-S* flip-flop illustrating its indeterminate output states (?) is shown below.

Truth Table

Input States		Output States	
S_t	R_t	Q_{t+1}	$\bar{Q}_{t+1}$
0	0	Q_t	$\bar{Q}_t$
1	0	1	0
0	1	0	1
1	1	?	?

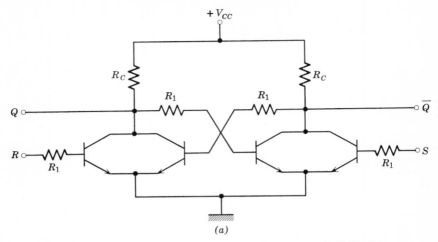

Figure 11.13 Schematic diagram of basic RTL reset-set (R-S) flip-flop.

A synchronous form of the basic storage circuit that has no indeterminate state in its truth table, illustrated below, is shown in Figure 11.14. In this

Truth Table

Input States		Output States	
S_t	R_t	Q_{t+1}	$\overline{Q}_{t+1}$
0	0	Q_t	$\overline{Q}_t$
1	0	1	0
0	1	0	1
1	1	$\overline{Q}_t$	Q_t

J-K flip-flop there are two identical three-input gates, which function as follows:

1. Gates Q_1 and Q_6 receive input signals from adjacent logic circuits.
2. Gates Q_3 and Q_4 sense the output signals of the flip-flop.
3. Gates Q_2 and Q_5 are driven by the clock inverter transistor Q_7 to prevent any response while the clock or trigger input signal to Q_7 is low.

These gates control the collector junction charge storage of Q_8 and Q_9 [26]. For the collector junction of either of these transistors to charge, the base must be at a high level relative to the collector. This requires that the output of the gate controlling the charge storage transistor be high, which occurs

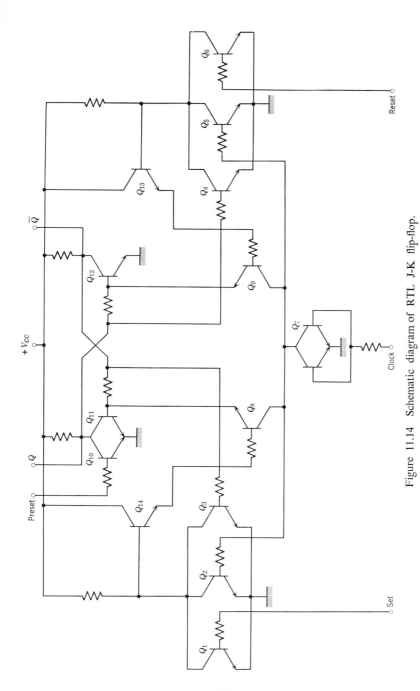

Figure 11.14 Schematic diagram of RTL J-K flip-flop.

only when all the gate inputs are low. The output flip-flop, Q_{11} and Q_{12}, is cross-connected to the gates, and since only one side is low at any time, either Q_8 or Q_9 charges, but not both, even when both logic input signals are low. The clock input is high during the charging time and is inverted to present a low level to the control gates. When the clock pulse falls, the collector of the charge storage transistor goes high, releasing its charge. If the associated output flip-flop transistor is off, the charge drives it on and thus changes the state of the output. Gate Q_{10} is used to preset the state of the flip-flop; Q_{13} and Q_{14} are emitter followers that deliver additional charging current to the charge storage transistors Q_8 and Q_9 to improve operating margins.

The circuit configurations in Figures 11.13 and 11.14 are especially adaptable to silicon monolithic integrated circuit technology since only transistors and resistors are involved [26].

11.4 TRANSISTOR RESISTOR LOGIC (TRL)

Perhaps the earliest logic system in which a single basic gate circuit was utilized is transistor resistor logic [27,28,29]. A basic TRL gate is illustrated in Figure 11.8b. The use of resistors as input gating elements was once attractive because of their relatively low cost and high reliability compared with diodes or transistors. In a monolithic integrated circuit the reverse is now true, especially when we consider the relatively tight tolerance requirements of TRL resistors. The brief discussion of TRL given in this section is intended mainly to add perspective to the overall discussion of micropower digital circuits.

The most negative feature of the TRL gate is the use of completely bilateral resistors as input gating elements. Resistors can be used for gating in this manner because the transistor input impedance is low enough to act as a clamp to isolate partially the separate inputs. However, resistor gating is still rather inefficient because the "1 state" or high-level input signal current of a given resistor can be seriously attenuated, before flowing into the transistor base, by shunt losses through the other $(M - 1)$ resistors in the gate fan-in whose inputs are at the "0 state" or low level. To minimize these shunt losses, the fan-in resistances must be large (or the total number of inputs must be small compared to RTL or DTL circuits). The use of large-input resistors markedly degrades the switching speed of the TRL circuit because (a) the time constants of the circuit increase, (b) it necessitates an increase in supply voltage and therefore in the amplitude of the signal voltage swing, and (c) there is little overdrive to switch the transistor either on or off.

The relatively large collector supply voltage required by a TRL gate raises its power consumption. The impact of reduced speed and increased power consumption is reflected by the TRL speed-power curve in Figure 11.12. The TRL circuits represented in this figure are optimum worst case designs equivalent to those previously discussed for RTL. The comparatively poor performance of the TRL gate strongly militates against its use in most micropower applications.

11.5 DIODE TRANSISTOR LOGIC (DTL)

The lack of isolation between inputs of the same gate exhibited by the TRL circuit can be corrected by replacing the input resistors with diodes as illustrated in Figure 11.8c [30,31]. For optimum worst case designs equivalent to those discussed for RTL, this DTL NOR circuit (assuming a positive signal level corresponds to a logical "1" while a nearly zero signal level corresponds to a logical "0") offers a speed-power performance improvement over the TRL gate in the micropower range, as Figure 11.12 indicates. Replacing the TRL gating resistors with diodes permits an improvement in switching speed because (a) the effective time constants of the circuit are reduced and (b) the collector supply voltage is reduced, thereby permitting a smaller signal voltage swing. The smaller supply voltage also contributes to a reduction in power consumption.

Perhaps the major shortcoming of the DTL NOR circuit is its lack of overdrive to switch the transistor both on and off. Overdrive current for the turn-on transient is limited by R_1, which is used to inhibit current hogging as described for the RTL gate. The worst

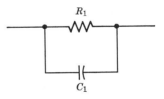

Figure 11.15 Offset level shifting network for DTL NAND R gate.

case current-hogging problem can be severe for DTL NOR gates because the tolerances of both an input diode and its associated inverter transistor can combine to promote current hogging. Turn-off overdrive is extremely limited since the high reverse impedance of the input diodes places the full burden of turn-off current on the negative supply voltage, $-V_{BB}$. The possibility of using slow reverse recovery input diodes to increase the turn-off base current is not feasible because a serious type of crosstalk would then occur.

A solution to the overdrive problems of the DTL NOR circuit is offered by the DTL NAND R circuit in Figure 11.8d [30,32,33]. By providing each transistor base with a separate turn-on path, through R_N and R_1, the current-hogging problem is circumvented and ample turn-on overdrive becomes available. A shunt capacitor C_1 can be used in parallel with R_1, as Figure 11.15

indicates. In particular, the turn-off base drive benefits from this feature because a driving transistor can "reach through" the input diodes and the level shifting network, R_1 and C_1, to draw reverse base current directly from the driven transistor. (Note that the use of a speed-up capacitor is prohibitive in the DTL NOR gate because its discharge path is blocked by the input diodes.) However, the primary purpose of R_1 is to provide a voltage offset in the base circuit to counterbalance the low-level input voltage to a cutoff gate and provide a stability margin ΔV_0. Consequently the addition of C_1 is somewhat conflicting since it brings with it a greater susceptibility to ac noise voltages at the input.

The use of a resistance R_1 and a capacitance C_1, as shown in Figure 11.15, is undesirable for monolithic integrated circuits. In such cases the functions of these elements can be performed by one or more diodes as illustrated for the DTL NAND D circuit in Figure 11.8e. Unlike the input diodes, however, whose reverse recovery time should be as small as possible, during turn-off the level shifting diodes ideally should not regain their high reverse impedance state until the transistor is cutoff. At milliampere transistor currents this requirement imposes limits on the effective minority carrier lifetime of the level shifting diodes [34]. In the microampere range of particular interest here, it is necessary that the average capacitance of the diodes, as seen by the transistor during the turn-off transient, be somewhat larger than the input capacitance of the transistor. Diode connected transistors [35] and N^+ emitter diffusions into P^+ isolation diffusions [36] can be used to achieve this behavior conveniently in monolithic structures.

For optimum worst case designs, the DTL NAND circuits offer markedly improved speed-power performance over the DTL NOR gate throughout the micropower range, as Figure 11.12 illustrates. The apparent slight inferiority of the NAND circuit with diode level shifting is a consequence of the severe worst case tolerances assumed for the level shifting diode forward voltage drops. Relaxing these assumptions in a realistic manner indicates that diode level shifting imposes no performance penalty. In both NAND circuits at $I_{CS} \leq 30\ \mu\text{a}$, a collector pull-up resistor R_C is needed to supply cutoff collector current I_{C0} and logic diode reverse currents in order to maintain the circuit performance cited in Figure 11.12. The saturated stability margin of the gate, ΔV_S, is improved by adding R_C. In addition, to maintain adequate cutoff stability margin ΔV_0, three-level shifting junctions become necessary for $I_{CS} \leq 30\ \mu\text{a}$ in the DTL NAND D because of the worst case tolerances in diode forward voltages. The performance displayed in Figure 11.12 indicates that, under the prevailing constraints, the DTL NAND gates do not suffer any overriding deficiencies in operating temperature range, fan-in, fan-out, tolerance immunity, and especially in stability margins for $10\ \mu\text{a} \leq I_{CS} \leq 1.0\ \text{mA}$.

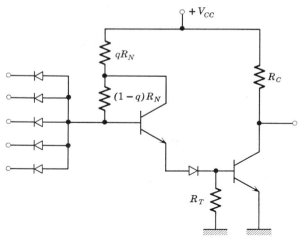

Figure 11.16 Improved form of basic DTL NAND D gate.

In order to achieve the best possible levels of performance from the DTL NAND D circuit, a negative supply voltage $-V_{BB}$ is necessary, as Figure 11.8 indicates. By replacing one of the level shifting diodes with an inverter base drive transistor [37], as indicated in Figure 11.16, the elimination of a negative supply voltage is possible without a performance penalty. The additional turn-on base drive that is available permits a reduction in the value of R_T which, in turn, maintains adequate turn-off base current with R_T grounded. The inverter base drive transistor reduces the current gain requirement of the output transistor for a given fan-out and permits wider resistor tolerances.

11.6 TRANSISTOR TRANSISTOR LOGIC (TTL)

Transistor transistor logic is a variation of DTL in which an input diode and a level shifting diode are replaced by a transistor [38] as illustrated in Figure 11.8f. In monolithic designs, a multiple emitter input transistor, as illustrated in Figure 11.17, is used for this circuit [36,39].

A disadvantage of the basic TTL gate is its cutoff stability margin, which is even smaller than that of the RTL gate because of the offset voltage of the coupling transistors. Level shifting diodes cannot be used conveniently as in the DTL NAND

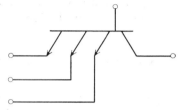

Figure 11.17 Multiple emitter input transistor for TTL.

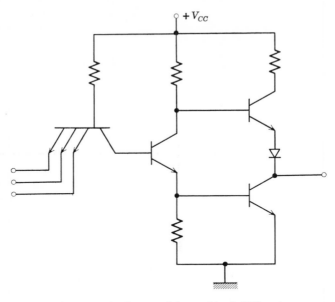

Figure 11.18   Improved form of basic TTL gate.

circuit, since they tend to block all base turn-off current. The collector pull-up resistor R_C improves the saturated stability margin of a basic TTL gate and is quite necessary in order to supply I_{C0} to inverter transistors and the reverse emitter currents of coupling transistors [8,40]. Without R_C, current robbing caused by unbalance in V_{BES} of inverter transistors driven by the same gate would be intolerable because of the inverse current gain of the coupling transistors. Because of coupling transistor inverse current gain, the reverse emitter input currents of a saturated TTL gate are considerably larger than the reverse diode input currents of a saturated DTL NAND gate. This tends to limit TTL at micropower levels.

For optimum worst case designs, the basic TTL circuit offers a speed-power performance slightly inferior to RTL, as Figure 11.12 indicates. To improve the inferior cutoff stability margin and the capacitive driving capability of TTL, the basic gate can be modified as indicated in Figure 11.18.

11.7   EMITTER COUPLED LOGIC (ECL)

An emitter coupled logic (ECL) gate designed expressly for micropower operation is illustrated in Figure 11.19 [41]. The similarity between this circuit and the improved TTL gate in Figure 11.18 is apparent. Also, we might regard the ECL gate as an advanced form of the RTL circuit in Figure 11.8b. The advantage in cutoff stability margin and driving capability is apparent.

A unique feature of this monolithic integrated circuit is that the gate voltage of the field effect resistors R_1 and R_2 (see Figure 1.17d) can be used to control dc power dissipation over a range in excess of three to one for $-16 \text{ V} \leq V_G \leq -8 \text{ V}$. This permits a limited speed-power trade-off following fabrication of the circuit. For a power drain of 60 μW, the propagation delay of this circuit is about 0.5 μsec with the output loaded by 50 pF to ground. The monolithic transistor emitter area is 0.5 × 0.4 mils with an emitter-base junction capacitance of 0.1 pF at zero bias.

A current steering ECL gate [42] that incorporates positive feedback to improve the stability margin or noise immunity is illustrated in Figure 11.20. In the usual emitter coupled logic gate [43] the base of Q_F is connected to a reference voltage source to provide a nonsaturating current steering circuit. In the ECL gate in Figure 11.20 the base of Q_F is connected to the common collectors of the input transistors so that the rising input voltage of Q_A, with Q_B and Q_C cutoff, for example, reduces the base voltage of Q_F and thereby provides positive feedback in steering the source current from Q_F to Q_A.

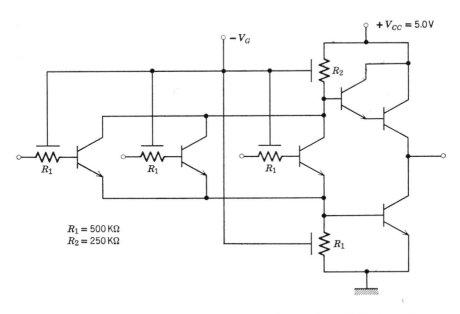

Figure 11.19 Emitter coupled logic (ECL) gate with insulated gate field effect resistors.

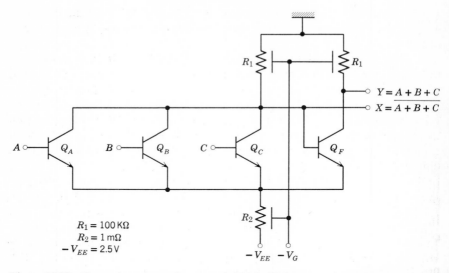

Figure 11.20 Current steering emitter coupled logic (ECL) gate with insulated gate field effect resistors and positive feedback.

The micropower ECL gate in Figure 11.20 exhibits the following salient features:

1. A minimum output voltage swing of 200 mV, which enhances operating speed.
2. Zero base input resistance R_1, which improves switching speed.

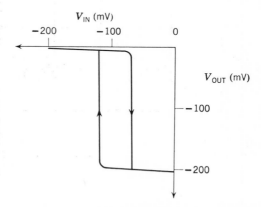

Figure 11.21 Hysteretic voltage transfer characteristic of current steering ECL gate.

3. A voltage transfer characteristic, as illustrated in Figure 11.21, with hysteresis resulting from the positive feedback which occurs during switching.

4. Improved noise immunity resulting from the hysteretic transfer characteristic.

5. Only one supply voltage.

6. Enhanced logic capability as a consequence of both OR ($Y = A + B + C$) and NOR ($X = \overline{A + B + C}$) outputs.

The average propagation delay of the current steering ECL gate in a five-stage ring oscillator, fabricated as a monolithic integrated circuit, is approximately 0.2 μsec for an average power dissipation of 10 μW and $V_G \simeq -11$ V. With the output loaded with 100 pF, the corresponding delay is about 8.6 μsec.

11.8 TUNNEL DEVICE LOGIC (TDL)

Tunnel diodes and tunnel rectifiers or backward diodes (see Section 1.5) are useful in some types of micropower digital circuits. The application of these devices, however, is limited by substantial obstacles:

1. The tunnel diode is a two-terminal switching element, and consequently the directivity of information flow in a logical network is a major consideration, whereas it is a relatively trivial matter in transistor circuits.

2. The reverse leakage current and breakdown voltage of a backward diode are quite poor compared with a conventional logic diode.

3. The fabrication processes for tunnel devices are largely incompatible with the technology of silicon monolithic integrated circuits.

Silicon backward diodes can be used in place of conventional diodes in DTL [44] as illustrated in Figure 11.22. Base current hogging is a problem in the NOR circuit. The low knee voltage of the backward diode in its forward direction eliminates the need for level shifting elements and a negative supply in the NAND configuration. The speed-power performance of the backward diode NAND circuit is approximately equal to that of RTL.

Micropower tunnel diode logic circuits [45] designed to operate in a monostable mode and perform the OR and AND logical functions are illustrated in Figure 11.23. In both these micropower circuits the tunnel diode is triggered from its low voltage to its high voltage state by the application of the proper

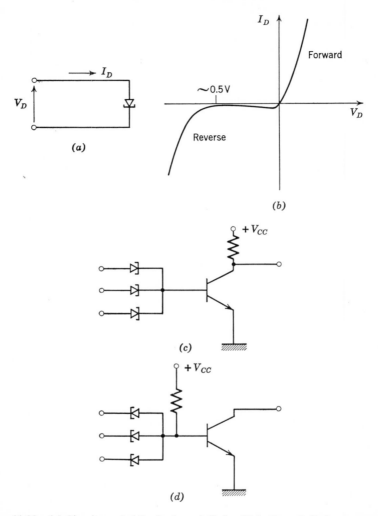

Figure 11.22 (a) Circuit symbol for backward diode; (b) backward diode current-voltage characteristic; (c) DTL NOR circuit using backward diodes; (d) DTL NAND circuit using backward diodes.

input signal currents. The output is in the form of a pulse followed by a recovery period. The OR gate is formed by backward diodes coupling multiple inputs to the gate, as shown in Figure 11.23a. The AND gate in Figure 11.23b is formed by adding current-limiting elements in series with the input backward diodes to control input current and improve tolerance immunity.

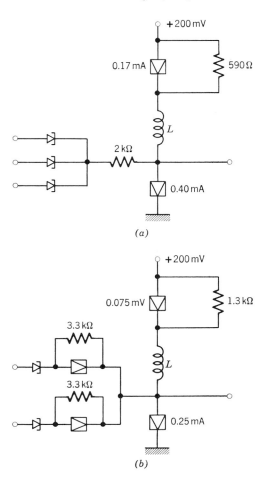

Figure 11.23 (a) Monostable tunnel diode OR gate; (b) monostable tunnel diode AND gate.

Low power operation is achieved by using 0.5 milliampere peak current germanium tunnel diodes and 200 mV supply voltages. This low supply voltage with a tunnel diode and resistor in parallel provides a constant current bias for the stage in its low state. The output pulse obtained is about 400 mV in amplitude, and the pulse width is determined by the value of the inductance L and the tunnel diode peak current. Such micropower tunnel device logic circuits have been constructed and operated at 10 megabit rates with 3 AND's and 1 OR in a sum-of-products gate dissipating 160 μW [45].

11.9 COMPLEMENTARY DIODE TRANSISTOR LOGIC (CDTL)

Although RTL, DTL, TTL, and ECL can perform quite well in the micropower range, a critical examination of their performance reveals unnecessary power dissipation. For instance, virtually the entire power consumption of a saturated RTL gate is wasted in the collector resistance R_C (Figure 11.8) and a substantial fraction of the power drain of a cutoff gate is wastefully dissipated in the same resistance. In a DTL NAND gate that is saturated, the dissipation in R_N and R_C is wasteful, as is the dissipation in R_N for a cutoff gate. The salient feature of complementary digital circuits [46] is the substitution of active transistor loads for passive resistor loads, such as R_C and R_N, in order to eliminate nonessential circuit power dissipation.

A complementary diode transistor logic (CDTL) gate that performs the NAND function is illustrated in Figure 11.24 [46–49]. Comparing this circuit with the DTL NAND gate in Figure 11.8d, it is evident that the collector resistor R_C has been replaced by a PNP transistor and two base circuit resistors, R_1 and R_T. When the gate (or Q_N) is saturated, Q_P is essentially cutoff. Consequently, in place of the power dissipated in R_C in a noncomplementary circuit, a much smaller dissipation in R_1 and R_T occurs in the complementary gate. When the gate (or Q_N) is cutoff, Q_P is turned on. Although this does not reduce the power dissipation of the complementary gate, since the DTL NAND gate supplies no static output current, the turn-off

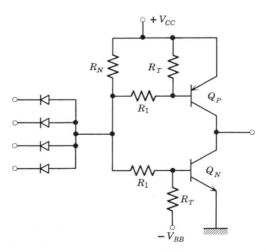

Figure 11.24 A complementary diode transistor logic (CDTL) gate. Typically, $I_{CS} = 10\ \mu$a, $R_1 = 1.6$ MΩ, $R_T = 2.4$ MΩ, $R_N = 680$ KΩ, $V_{CC} = 2.7$ V, $V_{BB} = 0$.

Complementary Diode Transistor Logic (CDTL)

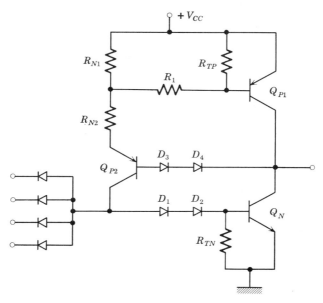

Figure 11.25 Improved CDTL gate.

transient response is improved as a result of the relatively large load capacitor charging current that Q_P can provide. The nonessential power dissipation in R_N is unchanged for the CDTL gate in Figure 11.24. With speed-up capacitors shunting R_1 (Figure 11.15), the reciprocal speed-power product of this circuit can be approximately twice that of an RTL gate [47] in the micropower range.

An improved CDTL gate suitable for fabrication as a monolithic integrated circuit is illustrated in Figure 11.25 [50,51]. When the gate is saturated, Q_{P1} is cutoff and (disregarding R_{TN}) Q_N and Q_{P2} will turn on as a PNPN device if

$$h_{FE}(Q_N) \cdot h_{FE}(Q_{P2}) > 1. \tag{11.62}$$

This is evident since the terminal current of the equivalent PNPN diode

$$I_E(Q_{P2}) = I_E(Q_N) = \frac{[I_{CBo}(Q_N) + I_{CBo}(Q_{P2})]h_{FE}(Q_N) \cdot h_{FE}(Q_{P2})}{1 - h_{FE}(Q_N) \cdot h_{FE}(Q_{P2})} \tag{11.63}$$

must be circuit limited for $h_{FE}(Q_N) \cdot h_{FE}(Q_{P2}) \to 1$. The principal added power saving in this gate occurs when it (or Q_N) is cutoff. In this state the base current of Q_{P2} must be zero, and thus the dissipation of the cutoff gate is

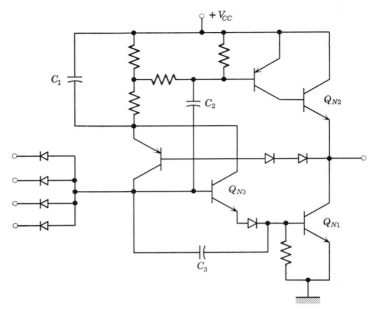

Figure 11.26 Additional speed-up elements added to improved CDTL gate.

virtually zero and fan-out is limited only by speed requirements. During the turn-off transient the base current of Q_{P2} flows through the load capacitance of the gate and thus permits Q_{P1} to supply a relatively large charging current to the load. The only current drain of this gate under static conditions is the emitter current of Q_{P2}, when the gate is saturated, which is

$$I_E(Q_{P2}) \approx \frac{V_{CC} - 3V_D}{R_{N1} + R_{N2}}, \tag{11.64}$$

where V_D is a diode forward voltage drop.

As Figure 11.26 illustrates, Q_{N3}, C_1, and C_3 can be added to the circuit in Figure 11.25 to reduce turn-on time while Q_{N2}, C_2, and C_3 reduce turn-off time. For a very similar circuit [51], with $V_{CC} = 4.0$ V the current drain of the gate is 35 μA when saturated and only 1 μA when cutoff. Turn-on time is 40 nanoseconds and turn-off time 60 nanoseconds for a 200 pF load capacitance.

The schematic diagram of a basic micropower complementary flip-flop without trigger circuitry is illustrated in Figure 11.27 [52,53,54]. The power-versus-frequency performance of this circuit is illustrated in Figure 11.28 [52]. The flatness of this curve for operating frequencies less than about 1.0 kHz shows that in this frequency range the small static or standby power drain is

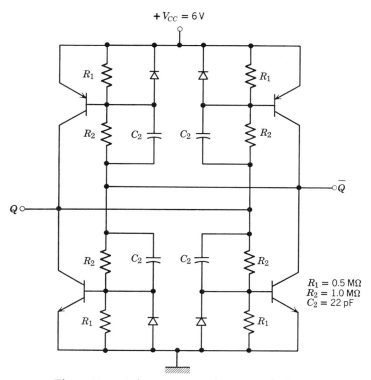

Figure 11.27 Micropower complementary flip-flop.

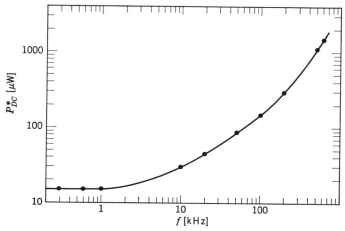

Figure 11.28 Power-versus-frequency profile of micropower complementary flip-flop (after Schmidt and Chace [52]).

the principal component of power dissipation. As the frequency of operation rises, however, the large increase in power consumption indicates that dynamic or transient power dissipation becomes much larger than the standby component. On this basis, the importance of including the transient power dissipation in calculations of the speed-power performance of complementary circuits is evident.

Figure 11.29 illustrates the schematic diagram of a monolithic micropower complementary flip-flop [55]. Here Q and $\bar{Q}$ are the true and false outputs, respectively; S_Q and $R_{\bar{Q}}$ are the trigger steering terminals that are used for a clocked set and reset in the clocked binary mode of operation; T is the clock or trigger input; and S_D and R_D are the direct set and reset inputs for

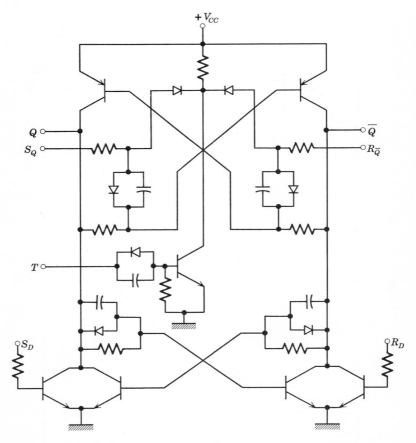

Figure 11.29 Monolithic micropower complementary flip-flop.

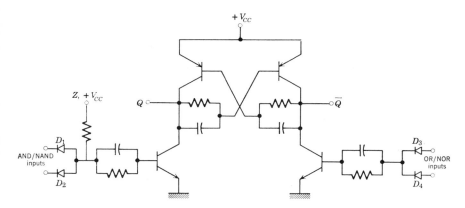

Figure 11.30 Universal micropower complementary logic element.

controlling the output state without using the trigger. The standby power drain of the circuit is about 500 μW for $V_{CC} = 3.0$ V.

A universal micropower logic element is illustrated in Figure 11.30 [52,56]. When the circuit is used as an AND or NAND gate, the logic terms are connected to the AND/NAND input diodes, with the complementary logic terms connected to the OR/NOR input diodes. The reverse connections are made for an OR or NOR gate. Although complementary input signals are required, the circuit provides these for use in subsequent levels of logic. A flip-flop is formed by removing point Z from V_{CC} and attaching it to Q while input $\bar{Q}$ is connected to an OR/NOR input. Circuits of this configuration can display power drains of less than 10 μW and switching times less than 0.5 μsec. Any standard logic form, instead of DTL as illustrated, can be used in the input section.

11.10 PULSE POWERED CIRCUITS (PPC)

A pulse powered circuit [57,58] operates in a mode in which the dc supply voltage of the circuit is gated on and off, preferably at a low-duty cycle, in order to reduce the average power drain of the circuit. In principle, both linear and digital circuits can operate in the pulse powered mode. Digital logic circuits can be operated in this mode by gating on their supply voltage during intervals when the circuits are called upon to process information and gating off their supply voltage during dormant intervals. Digital storage or memory circuits present another problem since a means must be found whereby a flip-flop which is reenergized will return to the same state in which it existed at the termination of the previous power pulse. To store the state of a flip-flop between pulses of power and to reestablish its state during a power pulse, capacitors can be used as temporary storage elements.

The basic RTL flip-flop illustrated in Figure 11.31a has a dc power drain

$$P_{DC}^* \simeq V_{CC}^2 \left[\frac{1}{R_C} + \frac{1}{R_1 + R_C} \right] \qquad (11.65)$$

for $V_{CC} \gg V_{BES} > V_{CES}$ and

$$P_{DC}^* \approx \frac{V_{CC}^2}{R_C} \qquad (11.66)$$

for $R_1 \gg R_C$. It is clear from previous discussion as well as (11.63) and (11.64) that R_C and R_1 must be large to permit small values of P_{DC}^*.

As illustrated in Figure 11.31b, the basic flip-flop is modified for pulse powered operation by addition of the ideal switches S_{CC}, S_1, and S_2 as well

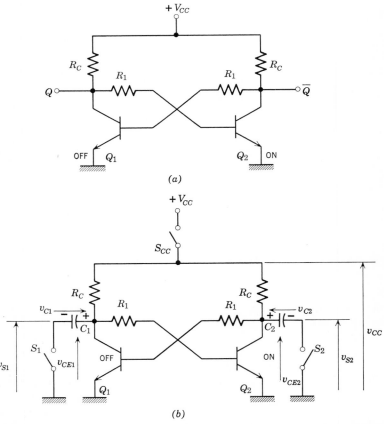

Figure 11.31 (a) Basic RTL flip-flop; (b) idealized configuration of a pulse powered flip-flop.

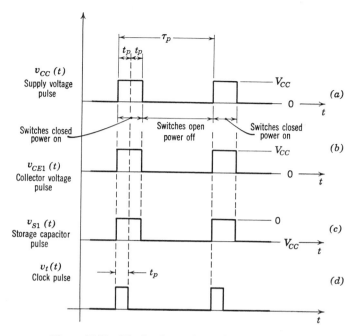

Figure 11.32 Idealized waveforms for a pulse powered flip-flop.

as capacitors C_1 and C_2. The switches are actuated so that the supply voltage waveform $v_{CC}(t)$ is as shown in Figure 11.32a. Consequently the charge stored on the capacitors, $Q_1 = C_1 V_{CC}$ and $Q_2 = C_2 V_{C2} = 0$, assures that each time the flip-flop is reenergized it will return to the same state in which it existed at the termination of the previous power pulse or to its "correct" state; that is, Q_2 will be saturated and Q_1 cutoff for the conditions in Figure 11.31b. Disregarding switching transient power dissipation, the average power drain is reduced by the factor $2t_p/\tau_p$ as a consequence of this pulse powered mode of operation of the flip-flop.

Under the condition that the flip-flop does not change state, as illustrated by the waveforms in Figure 11.32b,c, the average power drain is

$$P_{DC}^* \simeq V_{CC}^2 \left[\frac{1}{R_C} + \frac{1}{R_1 + R_C} \right] \frac{2t_p}{\tau_p} \qquad (11.67)$$

for $V_{CC} \gg V_{BES} > V_{CES}$ and

$$P_{DC}^* \approx \frac{V_{CC}^2}{R_C} \frac{2t_p}{\tau_p} \qquad (11.68)$$

for $R_1 \gg R_C$. If the flip-flop changes state every clock pulse,

$$P_{DC}^* \approx \frac{V_{CC}^2}{R_C} \frac{2t_p}{\tau_p} + CV_{CC}^2 \frac{1}{\tau_p}, \qquad (11.69)$$

for $C_1 = C_2 = C$ is the approximate average power drain. If ζ reflects the actual average rate of state change of a flip-flop compared to the clock rate,

$$P_{DC}^* \approx \frac{V_{CC}^2}{R_C} \frac{2t_p}{\tau_p} + CV_{CC}^2 \frac{\zeta}{\tau_p} \qquad (11.70)$$

for $0 < \zeta < 1$ describes the average power drain of the circuit. Typically, (11.70) is dominated by the first term, in which case comparing (11.66) and (11.70) indicates that P_{DC}^* can be reduced several orders of magnitude by controlling the value of t_p/τ_p (and not necessarily by increasing R_C).

The clock pulse width t_p as illustrated in Figure 11.32d must be sufficiently long to allow (a) the circuit to reach a reasonable steady state condition following the application of the power pulse and (b) the circuit to complete a change of state following the duration of the clock pulse. If we recognize that an interval of about three time constants, $3R_CC$, is required to charge a storage capacitor and that about an equal interval is required for a state transition [57], then

$$t_p \geq 3R_CC. \qquad (11.71)$$

The lower limit on τ_p is set by the leakage current of a switch S_1 or S_2 in the open position. For reliable circuit operation, the storage capacitor charge must not fall below about half its initial value during the interval the supply voltage is gated off. Consequently,

$$\tau_p \geq \tfrac{1}{2} CV_{CC} \frac{1}{I_L}, \qquad (11.72)$$

where I_L is the leakage current of an open switch. Combining (11.71) and (11.72) as equalities gives

$$\frac{2t_p}{\tau_p} = 12 \frac{I_L R_C}{V_{CC}}. \qquad (11.73)$$

For $I_L \sim 10^{-7}A$, $R_C \sim 10^4\,\Omega$, and $V_{CC} \sim 1.0$ V, (11.73) has a value $\sim 10^{-2}$ and for the less conservative case of $I_L \sim 10^{-9}A$, $R_C \sim 10^3\,\Omega$, and $V_{CC} \sim 10$ V, (11.73) is $\sim 10^{-6}$. Therefore it is clear that P_{DC}^* can be reduced markedly by pulse powered operation.

One method of implementing a pulse powered flip-flop is illustrated in Figure 11.33. In this circuit I_L is the reverse collector junction current of a PNP transistor. An arrangement whereby I_L becomes the gate leakage current of an insulated gate field effect transistor would reduce I_L substantially.

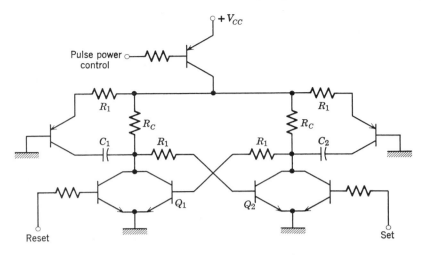

Figure 11.33 Practical configuration of a pulse powered flip-flop.

Figure 11.34 represents a multiuse technique by which additional power savings can be achieved in a pulse powered storage circuit. Several bits of information—two bits are illustrated—can be stored in association with a single flip-flop by using additional pairs of storage capacitors that are gated by multiphase control signals.

An error-correcting redundancy technique [59] that utilizes the unique features of a pulse powered flip-flop is illustrated in Figure 11.35. Assume that the internal capacitors, such as C_1 and C_2 in Figure 11.33, have stored some prior state of the flip-flop stages but one of the redundant triple is in error. Upon reapplication of power to the flip-flop, the capacitors C_{R1} through C_{R6} will cause a redistribution of capacitor charges so that the three circuits assume the state dictated by the majority capacitor charge.

In addition to reducing the power drain of flip-flops, the pulse powered mode of operation exhibits the following features:

1. The basic flip-flop itself need not be a high-impedance micropower design. In fact, widely used conventional milliwatt integrated circuits can be operated in a pulse powered mode by augmenting them with the proper external circuitry [59]. This opens the possibility of capitalizing on the improved noise immunity, operating temperature range, reliability, radiation resistance, cost, and availability of proven milliwatt circuit designs without the penalty of high power drain.

2. Improved noise immunity, since the circuits are deenergized a good deal of the time.

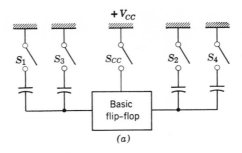

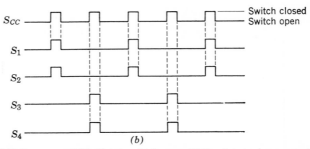

Figure 11.34 (a) Pulse powered flip-flop in which several bits of storage are associated with each bistable circuit; (b) multiphase control pattern for flip-flop.

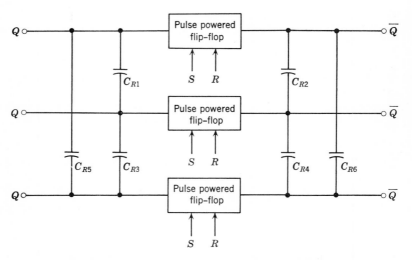

Figure 11.35 Triply redundant pulse powered flip-flops.

3. Improved system reliability through circuit redundancy and the multi-use technique described in connection with Figure 11.34.

REFERENCES

[1] A. W. Lo, *Introduction to Digital Electronics*, Addison-Wesley, Reading, Mass., 1967, Chapters 1 and 2.
[2] J. N. Harris et al., *Digital Transistor Circuits*, SEEC, Vol. 6, Wiley, New York, 1966.
[3] R. F. Shea et al., *Transistor Circuit Engineering*, Wiley, New York, 1957, Chapter 10.
[4] M. V. Joyce and K. K. Clarke, *Transistor Circuit Analysis*, Addison-Wesley, Reading, Mass., 1961, Chapters 10, 11, 12.
[5] L. P. Hunter, *Handbook of Semiconductor Electronics*, McGraw-Hill, New York, 1962, Section 15.
[6] A. I. Pressman, *Design of Transistorized Circuits for Digital Computers*, J. F. Rider, New York, 1959.
[7] Engineering Staff of Texas Instruments, Inc., *Transistor Circuit Design*, McGraw-Hill, New York, 1963, Chapters 27, 28, 29.
[8] J. D. Meindl et al., "Static and Dynamic Performance of Micropower Transistor Logic Circuits," *Proc. IEEE*, **52**, 1575–1580 (December 1964).
[9] R. P. Nanavati and R. A. Johnson, "Turn-on Delay Time and Its Prediction", *NEC*, **14**, 25–31 (October 1958).
[10] R. J. Wilfinger, "Predicting Transistor Turn-on Delay Time in the Common Emitter Configuration," *Solid State Design*, **3**, 34–41 (April 1962).
[11] J. A. Ekiss and C. D. Simmons, "Junction Transistor Transient Response Characterization," *Solid State J.*, **2**, 17–24 (January 1961).
[12] J. L. Moll, "Large Signal Transient Response of Junction Transistors," *Proc. IRE*, **42**, 1773–1783 (December 1954).
[13] J. W. Easley, "The Effect of Collector Capacity on the Transient Response of Junction Transistors," *IRE Trans. Electron. Devices*, **4**, 6–14 (January 1957).
[14] R. Beaufoy and J. Sparks, "The Junction Transistor as a Charge-Controlled Device," *J. Aut. Tel-Elec.* (London), **13**, 310–327 (October 1957).
[15] H. J. Kuno, "Rise and Fall Time Calculations of Junction Transistors," *IEEE Trans. Electron. Devices*, **11**, 151–155 (April 1964).
[16] Y. Hsia and F. Wang, "Switching Waveform Prediction of a Simple Transistor Inverter Circuit," *IEEE Trans. Electron. Devices*, **12**, 626–631 (December 1965).
[17] J. G. Linvill, "Lumped Models of Transistors and Diodes," *Proc. IRE*, **46**, 1141–1152 (June 1958).
[18] D. J. Hamilton et al., "Comparison of Large Signal Models for Junction Transistors," *Proc. IEEE*, **52**, 239–248 (March 1964).
[19] R. P. Nanavati, "Prediction of Storage Time in Junction Transistors," *IRE Trans. Electron. Devices*, **7**, 9–15 (January 1960).
[20] S. Steiner, "Turn-off Transistion Mechanism," *Solid State Design*, **3**, 29–34 (June 1962).
[21] R. H. Beter et al., "Surface-barrier Transistor Switching Circuits," *1955 IRE Conv. Record*, **4**, Pt. 4, 139–145.
[22] J. W. Easley, "Transistor Characteristics for Direct Coupled Transistor Logic Circuits," *IRE Trans. Electron. Computers*, **7**, 6–16 (March 1958).

[23] J. R. Harris, "Direct-Coupled Transistor Logic Circuitry," *IRE Trans. Electron Computers*, **7**, 2–6 (March 1958).

[24] A. T. Watts, "A Microminiature Digital Integrator Using Micropower Circuits," in *Micropower Electronics*, E. Keonjian, Ed., Macmillan, New York, 1964, pp. 85–104.

[25] W. W. Gaertner, "Micropower Microelectronic Subsystems," in *Micropower Electronics*, E. Keonjian, Ed., Macmillan, New York, 1964, pp. 57–84.

[26] D. C. Davies et al., "An Integrated Charge-Control J-K Flip-Flop," *IEEE Trans. Electron. Devices*, **11**, 556–562 (December 1964).

[27] W. D. Rowe, "The Transistor NOR Circuit," 1957 *IRE WESCON Conv. Record*, Pt. 4, 231–245 (August 20–23, 1957).

[28] W. D. Rowe and G. H. Royer, "Transistor NOR Circuit Design," *Trans. AIEE Commun. and Electron.*), **76**, 263–267 (July 1957).

[29] Q. W. Simkins, "Transistor Resistor Logic," *Semicond. Prod.*, **2**, 34–38 (April 1959).

[30] W. B. Cagle and W. H. Chen, "A New Method of Designing Low Level, High Speed Logic Cricuits," *WESCON Conv. Resord*, Pt. 2, **1**, 3–9 (August 1957).

[31] P. W. Becker, "Static Design of Transistor Diode Logic," *IRE Trans. Circuit Theory*, **9**, 461–476 (December 1961).

[32] E. G. Rupprecht et al., "Hyperfast Diffused-Silicon Diode and Transistor for Logic Circuits," *Inter. Solid-State Circuits Conf., Dig. Tech. Papers*, Vol. II, 72–73 (February 1959).

[33] D. P. Masher, "The Design of Diode Transistor NOR Circuits," *IRE Trans. Electron. Computers*, **9**, 15–25 (March 1960).

[34] W. M. Hailey, "Diode Transistor Minority Carrier Lifetime Relationships in Integrated Diode Transistor Logic Circuit," *Solid State Design*, **5**, 21–26 (June 1965).

[35] H. C. Lin, "Diode Operation of a Transistor in Functional Blocks," *IEEE Trans. Electron. Devices*, **10**, 189–194 (May 1963).

[36] L. D. Hirsch et al., "Two Approaches to Charge Control in Saturated Logic Gates," *1966 Inter. Solid-State Circuits Conf., Dig. Tech. Papers*, Vol. IX, 14–15 (February 1966).

[37] R. Bohn and R. Seeds, "Collector Tap Improves Logic Gating," *Electron. Design*, **12**, 48–55 (August 3, 1964).

[38] R. H. Beeson and H. W. Ruegg, "New Forms of All Transistor Logic," *Inter. Solid-State Circuits Conf. Dig. Tech. Papers*, Vol. V, 10–11 (February 1962).

[39] R. Bohn et al., "50 MC Monolithic Integrated Circuits for Digital Application," *Solid State Design*, **4**, 25–32 (July 1963).

[40] H. C. Josephs, "The Effect of Fan-in and the Pull-up Resistor on Margins in TTL," *Proc. IEEE*, **53**, 532–533 (May 1965).

[41] J. E. Price, "A Compatible MOS-Bipolar Device Technology for Low Power Integrated Circuits," International Electron. Devices Meeting, October 26–28, 1966, Washington, D.C.

[42] C. A. Bittman et al., "Micropower Functional Electronic Blocks," Technical Report AFAL-TR-66-338, 58–82 (September 1966).

[43] J. A. Narud and C. S. Meyer, "Characterization of Integrated Logic Circuits," *Proc. IEEE*, **52**, 1551–1564 (December 1964).

[44] D. F. Allison et al., "KMC Planar Transistors in Microwatt Logic Circuitry," *Inter. Solid State Circuits Conf., Dig. Tech. Papers* (February 1961).

[45] P. Gardner et al., "Application of Tunnel Diodes to Micropower Logic Circuits," *Solid State Electronics*, **7**, 747–751 (1964).

[46] R. H. Baker, "Maximum Efficiency Transistor Switching Circuits," MIT Lincoln Laboratory Report TR-110 (March 22, 1956).

References

[47] D. G. Patterson, "Micropower Complementary Logic," *Proc. IEEE*, **52**, 1581–1583 (December 1964).

[48] R. A. Tietsch, "Complementary Microwatt Logic Circuits," *Electron. Equip. Eng.*, Vol. II, 50–52 (August 1963).

[49] J. C. Sturman, "Micropower Transistor Logic Circuits," NASA Technical Note D-1462 (February 1963).

[50] R. Y. Hung and H. C. Lin, "Integrated Complementary Logic Block for Low Power Dissipation," *Proc. NEC*, Chicago, Ill., **20**, 163–168 (October 1964).

[51] R. Y. Hung and H. C. Lin, "Integrated High Speed Low Power Complementary Bipolar Transistor Nand Gate," Session 1, WESCON, San Francisco, Calif. (August 24–27, 1965).

[52] W. G. Schmidt and D. E. Chace, "Design Aspects of Minimal-Power Digital Circuitry," M.I.T. Lincoln Laboratory Group Report 1965–1966 (February 9, 1965).

[53] D. E. Chace, "A High Low-Power Complementary Flip-Flop," *Electron. Commun.*, **1**, 5 (November/December 1966).

[54] J. Schroeder, "High Speed Complementary Flip-Flop Features Extremely Low Power Dissipation," Motorola Semiconductor Products, Inc., Application Note AN-229.

[55] G. Y. Chang, "A Silicon Monolithic Micropower Complementary Flip-Flop," Session 1, WESCON, San Francisco, Calif., **9**, 24–27 (August 1965).

[56] W. G. Schmidt, "A Universal Micropower Logic Element," *Proc. IEEE*, **53**, 629 (June 1965).

[57] R. H. Baker et al., "Pulse Powered Circuits," Technical Report TR 65-1, Center for Space Research, M.I.T., Cambridge, Mass.

[58] R. H. Baker et al., "Pulse Powered Circuits," *IEEE Trans. Electron. Computers*, **15**, 321–323 (June 1966).

[59] R. E. McMahon and N. Childs, "Micropower Redundant Circuits Correct Errors Automatically," *Electronics*, **40**, 66–69 (February 6, 1967).

Chapter 12

Field Effect Transistor Digital Circuits

Digital circuits that utilize insulated gate field effect transistors can be designed to provide excellent micropower performance. With few exceptions, they have been implemented most effectively as silicon monolithic integrated circuits, which are described as metal-oxide semiconductor (MOS) integrated circuits. In fact, the small silicon area required by a typical insulated gate field effect transistor (IGFET) integrated circuit, compared with its bipolar transistor counterpart, permits the highest levels of functional complexity or large-scale integration. The purpose of this chapter is to consider the salient aspects of IGFET digital circuits for operation in the micropower range.

12.1 THE FET INVERTER

The most elementary circuit configuration that incorporates the major features of FET digital circuits is the inverter in Figure 12.1. Several variations of this simple circuit are discussed in this section [1,2].

Linear Resistor Loads [1]

The static drain characteristics and transfer characteristic of an n-channel enhancement IGFET are illustrated in Figure 12.2. The load line for an inverter with a linear external drain resistance, R_D, is illustrated in Figure 12.2a. The circuit voltage transfer characteristic is shown in Figure 12.3.

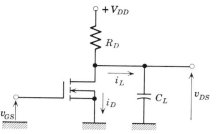

Figure 12.1 n-Channel enhancement insulated gate field effect transistor inverter with a linear load resistance.

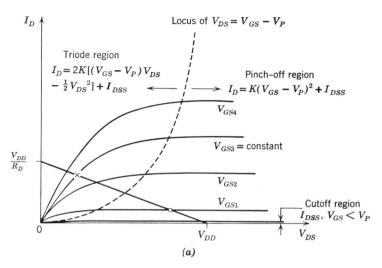

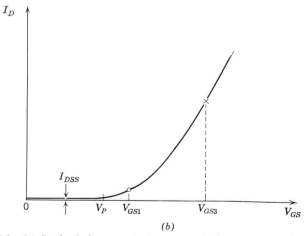

Figure 12.2 (a) Static drain current, I_D, versus drain-to-source voltages, V_{DS}, for an n-channel enhancement IGFET; (b) transconductance transfer characteristic for an n-channel enhancement IGFET.

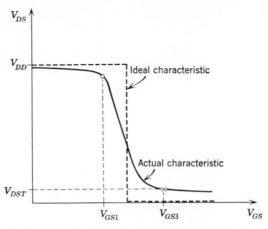

Figure 12.3 Transfer function of an IGFET inverter with a linear resistor load.

When the inverter is cutoff,

$$V_{GS} < V_P, \tag{12.1a}$$

$$I_D = I_{DSS} \simeq 0, \tag{12.1b}$$

and

$$V_{DS} \simeq V_{DD} > V_{GS} - V_P \tag{12.1c}$$

describe its behavior. In the active region where inverter voltage gain is large

$$V_{GS} > V_P \tag{12.2a}$$

from (1.38) and

$$I_D = K(V_{GS} - V_P)^2 + I_{DSS} \tag{12.2b}$$

$$V_{DS} = V_{DD} - I_D R_D > V_{GS} - V_P. \tag{12.2c}$$

Finally, in the triode region

$$V_{GS} > V_P \tag{12.3a}$$

$$I_D = 2K[(V_{GS} - V_P)V_{DS} - \tfrac{1}{2}V_{DS}^2] + I_{DSS} \tag{12.3b}$$

from (1.36) and

$$V_{DS} = V_{DD} - I_D R_D < V_{GS} - V_P. \tag{12.3c}$$

Typically, $V_{DS} \ll V_{DD}$ in this region so that

$$I_D \simeq \frac{V_{DD}}{R_D} \gg I_{DSS} \tag{12.4}$$

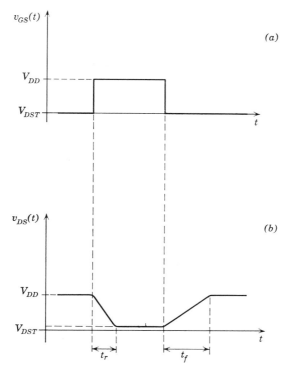

Figure 12.4 (a) Input voltage waveform; (b) output voltage waveform for the IGFET inverter.

is controlled largely by R_D and

$$V_{DS}^2 - 2(V_{GS} - V_P)V_{DS} + \frac{1}{K}\frac{V_{DD}}{R_D} = 0 \qquad (12.5)$$

gives V_{DST} in Figure 12.3.

The average power consumption of the inverter is

$$P_{DC}^* \simeq \frac{1}{2}\frac{V_{DD}^2}{R_D}, \qquad (12.6)$$

which can be reduced to micropower levels by increasing R_D.

If the inverter is excited with a step input voltage as described by Figure 12.4a, a turn-on delay time is avoided since the average gate-to-source capacitance C_{GS} (see Figure 1.12) is fully charged instantaneously. For an equivalent load capacitance $C_L \gg C_{GD}$ the drain node equation

$$-C_L \frac{dv_{DS}}{dt} + \frac{V_{DD} - v_{DS}}{R_D} = i_D \tag{12.7}$$

describes $v_{DS}(t)$ during the rise time t_r. Since $v_{DS} \leq V_{DD} = v_{GS} \simeq v_{GS} - V_P$ and $V_{DD} \gg V_P$, i_D is described approximately during t_r by (12.3b) or

$$I_D \simeq 2K \left(\frac{V_{DD} - V_P}{V_{DD}}\right)^2 [V_{DD} V_{DS} - \tfrac{1}{2} V_{DS}^2]. \tag{12.8}$$

Combining and solving (12.7) and (12.8) yields [1]

$$v_{DS}(t) = V_{DD} \frac{\sqrt{\alpha^2 + 1} - (\alpha - 1)\tanh[(\sqrt{\alpha^2 + 1}/\alpha)\tau]}{\sqrt{\alpha^2 + 1} + \tanh[(\sqrt{\alpha^2 + 1}/\alpha)\tau]} \tag{12.9a}$$

or

$$v_{DS}(t) = V_{DD}\left(1 - \frac{\alpha \tanh \tau}{\alpha + \tanh \tau}\right) \tag{12.9b}$$

for the typical condition $\alpha^2 \gg 1$ with

$$\alpha = 2KR_D V_{DD}\left(1 - \frac{V_P}{V_{DD}}\right)^2 \tag{12.10a}$$

and

$$\tau = \frac{\alpha t}{2 R_D C_L} = K \frac{V_{DD}}{C_L}\left(1 - \frac{V_P}{V_{DD}}\right)^2 t. \tag{12.10b}$$

From (12.10), the normalized rise time is

$$\tau_r = \tanh^{-1}\left(\frac{80\alpha}{91\alpha + 9}\right) \simeq 1.37, \tag{12.11}$$

and from this

$$t_r = \frac{1.37}{K} \frac{C_L}{V_{DD}} \frac{1}{(1 - V_P/V_{DD})^2} \tag{12.12}$$

gives the inverter rise time for

$$0.9(V_{DD} - V_{DST}) + V_{DST} \leq v_{DS}(t) \leq 0.1(V_{DD} - V_{DST}) + V_{DST} \tag{12.13}$$

At the outset of the fall time t_f, the IGFET is cutoff instantaneously so that the turn-off transient is a simple $R_D C_L$ charging function given by

$$v_{DS}(t) = V_{DD}\left[\left(\frac{V_{DST}}{V_{DD}} - 1\right)\epsilon^{-t/R_D C_L} + 1\right]. \tag{12.14}$$

The FET Inverter

From (12.14), the normalized fall time corresponding to the limits of (12.13) is

$$\tau_f = 1.10\alpha \qquad (12.15a)$$

and on this basis the inverter fall time is

$$t_f \simeq 2.2 R_D C_L \qquad (12.15b)$$

for

$$V_{DST} \simeq V_{DD} \frac{1}{1+\alpha} \qquad (12.16a)$$

or

$$V_{DST} \simeq \frac{V_{DD}}{1 + 2KR_D V_{DD}(1 - V_P/V_{DD})^2} \qquad (12.16b)$$

from (12.10) as $\tau \to \infty$.

The total switching time of the inverter is

$$t_{sw} = t_r + t_f \qquad (12.17a)$$

$$= \frac{1.37}{K} \frac{C_L}{V_{DD}} \frac{1}{(1 - V_P/V_{DD})^2} + 2.2 R_D C_L, \qquad (12.17b)$$

where from (1.37)

$$K = \frac{\mu_0 \epsilon_{0x} W_C}{2 T_{0x} L_C} \qquad (12.18)$$

describes physical properties of the IGFET. From Figure 12.5 it is evident that the current i_{Lr} available for discharging C_L during t_r is larger than the charging current i_{Lf} of C_L during t_f. Consequently t_f tends to dominate t_{sw} in the micropower range where R_D becomes quite large.

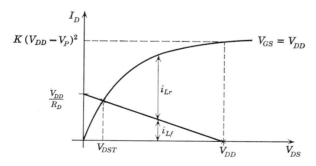

Figure 12.5 Load line diagram for IGFET inverter illustrating current available for discharging, i_{Lr}, and charging, i_{Lf}, load capacitance.

Depletion Transistor Loads

As Figure 1.13 illustrates, IGFET's can be designed to operate both as enhancement devices and as depletion devices [1]. Figure 12.6 describes a technique for utilizing an n-type depletion IGFET as the nonlinear load resistance for an n-type enhancement device. There are several advantages of this arrangement:

1. The transient currents available for discharging and charging C_L are more nearly equal, as Figure 12.6b indicates.
2. The voltage transfer characteristic of the inverter is more nearly ideal and the circuit is generally more stable.

However, the fabrication of both depletion and enhancement transistors in a monolithic integrated circuit entails added process complexity.

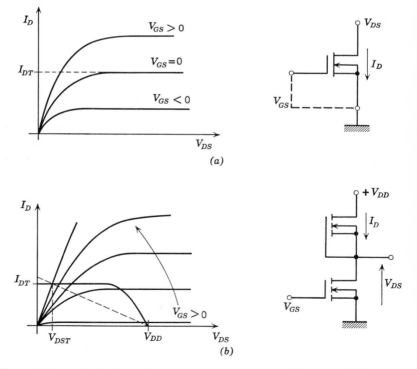

Figure 12.6 (a) Static drain characteristics for an n-channel depletion IGFET; (b) load line diagram for IGFET inverter using an n-channel depletion transistor load.

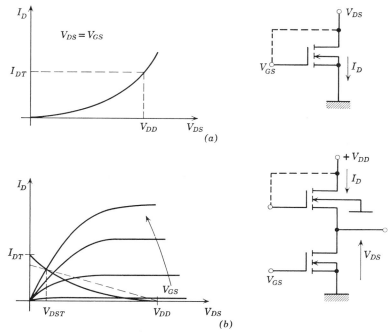

Figure 12.7 (a) Static drain characteristic for an *n*-channel enhancement IGFET with $V_{DS} = V_{GS}$ or $V_{GD} = 0$; (b) loadline diagram for IGFET inverter using an *n*-channel enhancement transistor load.

Enhancement Transistor Loads [3–9]

From Figure 1.14, it is evident that an IGFET, in an array of similar devices fabricated as a monolithic integrated circuit, is a self-isolating device. That is, separate isolation junctions are unnecessary since the source and drain regions themselves never become forward biased and therefore always remain isolated from the substrate. This self-isolation feature is a major advantage in increasing the number of devices per unit area and therefore reducing the cost of MOS integrated circuits. It is important that the load resistor of an IGFET inverter does not compromise this advantage.

As already discussed, depletion transistors add process complexity. Of the remaining types of resistors (Figure 1.17), the enhancement transistor itself is a leading candidate because (a) it entails no additional process complexity whatsoever, (b) it is self-isolating, (c) it occupies a relatively small surface area because of its large sheet resistivity, and (d) its temperature coefficient is advantageously matched to that of the inverter transistor.

Figure 12.7 illustrates the use of an enhancement transistor as a nonlinear load resistor. Although this configuration is based on extremely favorable fabrication requirements, its electrical behavior in terms of switching speed, transfer characteristic curve, and overall stability of the design is the poorest of the four inverters discussed in this section.

It is important to recognize that the ratio of channel width to length, W_c/L_c, in essence the transconductance of the load transistor, must be much smaller than that of the inverter transistor, so that $V_{DST} \ll V_{DD}$. From (1.42),

$$\frac{1}{\partial I_D/\partial V_{DS}} = r_{ds} = \frac{1}{2K[(V_{GS} - V_{DS}) - V_P]} \quad (12.19)$$

gives the drain-to-source resistance of an IGFET in the triode region or its low resistance state. Since the drain characteristics are nearly linear in this region and $V_{DS} \to 0$,

$$r_{ds} \simeq \frac{1}{2K[V_{GS} - V_P]}, \quad (12.20a)$$

which gives

$$r_{ds} = \frac{1}{g_m} = \left(\frac{\partial I_D}{\partial V_{GS}}\right)^{-1} \quad (12.20b)$$

from (1.39). Since

$$g_m = 2K(V_{GS} - V_P) \quad (12.21a)$$

$$= 2\left(\frac{\mu_c \epsilon_{0x} W_c}{2T_{0x} L_c}\right)(V_{GS} - V_P) \quad (12.21b)$$

from (1.37), it is evident that W_c/L_c for the load transistor must be relatively small for preservation of two discrete logical levels.

Complementary Transistor Loads [1,10–18]

From the standpoint of electrical performance, the most powerful IGFET inverter is the complementary circuit that uses the n- and p-channel enhancement devices illustrated in Figure 12.8. As the load line curves in Figure

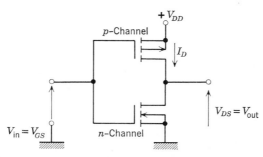

Figure 12.8 Schematic diagram for complementary IGFET inverter using a p-channel enhancement transistor load.

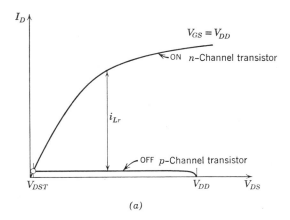

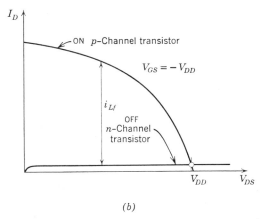

Figure 12.9 Load line diagram for complementary IGFET inverter in (a) ON state and (b) OFF state.

12.9a, b indicate, the values of the "0" and "1" logical levels in this circuit are essentially ideal. For $V_{in} \simeq V_{DD}$, the n-channel device is ON and the p-channel device is OFF so that $V_{out} \simeq V_{DD}$. This circuit is virtually ideal from a micropower standpoint since its standby current drain is essentially zero (that is, the leakage current of an OFF IGFET which is typically in the sub-microampere range) in either state. Furthermore, large and equal (for matched p- and n-type devices) currents, i_{Lr} and i_{Lf}, are available for rapid discharging and charging of capacitive loads presented by similar inverters.

The static transfer characteristic curves of the four inverters considered in

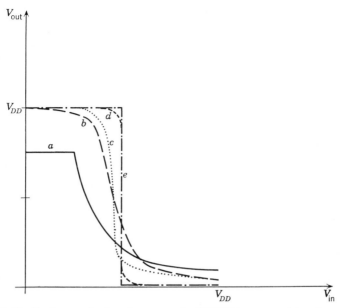

Figure 12.10 Static transfer function curves for IGFET inverters: (*a*) *n*-channel enhancement IGFET load; (*b*) linear resistor load; (*c*) *n*-channel depletion IGFET load; (*d*) *p*-channel enhancement IGFET load (complementary circuit); and (*e*) ideal transfer function.

this section, as well as an ideal curve, are compared in Figure 12.10. In general, the degree of ideality of a transfer function curve corresponds to the degree of difficulty encountered in fabricating the associated inverter as a monolithic integrated circuit. The complementary inverter (curve *d*) is most nearly ideal electrically, considering both static and dynamic behavior, but the most demanding from the standpoint of fabrication. The all-enhancement transistor inverter (curve *a*) is the poorest electrically, considering both static and dynamic behavior, but the most attractive for monolithic fabrication. A cross section of a monolithic complementary pair of transistors is illustrated in Figure 12.11*a* and a monolithic *n*-type enhancement inverter in Figure 12.11*b*.

For the voltage step driving function illustrated in Figure 12.4*a* with $V_{DST} \simeq 0$, during turn-on of the complementary inverter in Figure 12.8 the *n*-channel device conducts and the *p*-channel unit is cutoff; the reverse is true during turn-off. Consequently the rise time transient is a special case of the analyses presented for the circuit in Figure 12.1 in which $R_D \to \infty$ [1]. Because of the symmetry of the circuit, the turn-off or fall time transient must have the same shape as the rise time. Thus, from (12.9*b*) during the

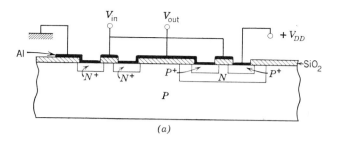

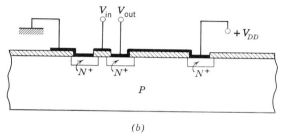

Figure 12.11 (a) Cross section of monolithic complementary pair inverter; (b) cross section of monolithic n-channel enhancement inverter.

rise time
$$v_{DS}(t) = v_{\text{out}}(t) = V_{DD}(1 - \tanh \tau), \tag{12.22}$$
$V_{DD} \geqslant v_{\text{out}}(t) \geqslant 0$, and during the fall time
$$v_{DS}(t) = v_{\text{out}}(t) = V_{DD} \tanh \tau \tag{12.23}$$
for matched n- and p-channel units. The normalized 10 to 90% rise and fall times are
$$\tau_r = \tau_f = 1.37, \tag{12.24a}$$
which yield
$$t_r = t_f \simeq \frac{1.37}{K} \frac{C_L}{V_{DD}} \frac{1}{(1 - V_P/V_{DD})^2}. \tag{12.24b}$$

The total energy dissipated in charging and discharging a capacitor C_L to a voltage V_{DD} is given by
$$E = C_L V_{DD}^2, \tag{12.25}$$
so that the approximate average power drain of the complementary inverter is
$$P_{DC}^* = C_L V_{DD}^2 f_0 + V_{DD} I_{DSS}, \tag{12.26}$$

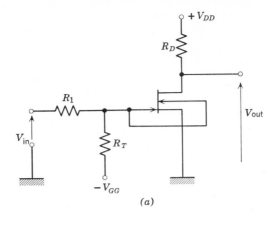

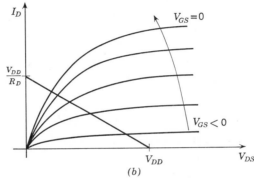

Figure 12.12 (a) Schematic diagram for junction gate FET inverter; (b) inverter load line diagram.

where f_0 is the operating frequency and $V_{DD}I_{DSS}$ is the standby power drain of the circuit.

Junction Gate Field Effect Transistor Inverters [19]

Junction gate field effect transistors also can be used in inverter circuits, as Figure 12.12 illustrates. However, since the junction gate device is inherently a depletion unit, it requires a turn-off supply voltage, such as $-V_{GG}$ in Figure 12.12, a turn-off resistor R_T, and a level shifting element, such as R_1, in order to cutoff the inverter. This additional complexity, compared with enhancement IGFET inverters, is a pronounced disadvantage in most digital applications. In addition, the JGFET is not self-isolating in monolithic integrated circuits, as Figure 1.11 suggests.

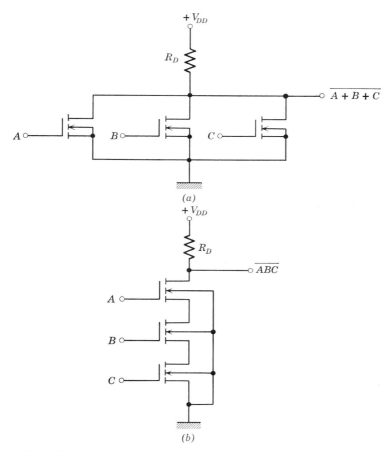

Figure 12.13 (a) Parallel IGFET NOR gate; (b) series IGFET NAND gate.

12.2 LOGIC CIRCUITS

The simple inverter is the most fundamental FET logic circuit. It serves as the basis for composite circuit structures that provide the AND, OR, and data storage functions.

If a logical one is defined to correspond to the high voltage level and a logical zero to the low voltage level, the circuit in Figure 12.13a functions as a three-input NOR gate and the circuit in Figure 12.13b as a three-input NAND gate. For the worst case configuration of NOR gates [4] illustrated in Figure 12.14,

$$\bar{V}_{DST} \leq \underline{V}_P - \Delta V_0 \qquad (12.27)$$

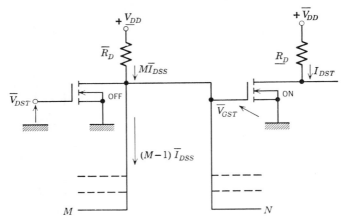

Figure 12.14 Worst case circuit configuration for parallel IGFET NOR gates.

defines the condition for sustaining a cutoff gate whose input voltage is $\bar{V}_{DST}$, threshold or pinchoff voltage is $\underline{V}_P$, and turn-off stability margin or noise margin is ΔV_0. Then

$$\bar{R}_D \leq \frac{\bar{V}_{DD} - \bar{V}_{GST} - \Delta V_S}{M \bar{I}_{DSS}} \tag{12.28}$$

defines the condition on $\bar{R}_D$ for sustaining the gate-to-source voltage $\bar{V}_{GST}$ at its turn-on value despite reductions caused by leakage current $\bar{I}_{DSS}$ and noise ΔV_S, and

$$\underline{R}_D \geq \frac{\bar{V}_{DD}}{I_{DST}} \tag{12.29}$$

defines the condition on $\underline{R}_D$ for preventing the drain current I_{DST} from exceeding the value at which $\bar{V}_{GST}$ can maintain the drain-to-source voltage in accordance with (12.27).

Figure 12.15 illustrates two complementary logic gates. For an iterative chain of these gates with a fan-out and fan-in of one and an output node capacitance C_L, the average propagation delay time is [12,13]

$$t_{pd} \approx 0.45 \frac{C_L}{V_{DD}} \left[\frac{1}{K_n(1-v_n)^2} + \frac{1}{K_p(1-v_p)^2} \right] \tag{12.30}$$

for

$$v_n + v_p < 1.0, \qquad \frac{K_p}{K_n} \geq 0.2,$$

$$v_n = \frac{V_{pn}}{V_{DD}}, \qquad v_p = \frac{|V_{pp}|}{V_{DD}},$$

$$K_n = \frac{\mu_{0n}\epsilon_{0xn}W_{cn}}{2T_{0xn}L_{cn}} \quad \text{and} \quad K_p = \frac{\mu_{0p}\epsilon_{0xp}W_{cp}}{2T_{0xp}L_{cp}},$$

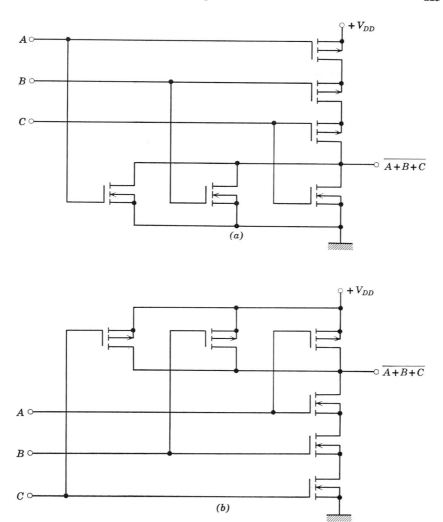

Figure 12.15 (a) Parallel n-channel IGFET complementary NOR gate; (b) series n-channel IGFET complementary NAND gate.

where the subscript n pertains to n-channel devices and p to p-channel devices. For multiple input logic gates, (12.30) can be generalized by using effective values for K_n and K_p. For series units the effective value of K is K/n and for parallel units it is Km where n and m are the number of devices being switched simultaneously.

The complementary IGFET logic gate offers outstanding advantages for micropower operation: (a) its standby power drain is extremely small, (b) its operating speed is better than other IGFET gates and good in comparison to typical transistor gates, (c) its speed-power performance is quite outstanding in the micropower range, (d) its noise immunity is excellent, (e) its fan-in and fan-out capability is large, (f) it operates from a single supply voltage and is highly tolerant of supply voltage variations, (g) circuit configurations are simple and inherently immune to wide tolerances in device parameters, and (h) its small area requirement as a monolithic integrated circuit and tolerance immunity promote high yield and low cost.

12.3 STORAGE CIRCUITS

A simple n-channel IGFET set-reset flip-flop formed by combining a pair of two-input NOR gates (Figure 12.13a) is illustrated in Figure 12.16. The flip-flop is ON or output Q is at logical "1" when Q_3 is conducting and all other transistors are cutoff. A positive voltage at the R input is used to reset the flip-flop to the 0 or OFF state where Q_2 is conducting and all other transistors are cutoff.

A complementary set-reset flip-flop is illustrated in Figure 12.17 [13,14,20]. The total power dissipated in this circuit is

$$P_{DC}^* = 2C_L V_{DD}^2 f_0 + P_S, \qquad (12.31)$$

where the first term represents switching or transient power and the second standby power. The total output node capacitance is C_L. Figure 12.18 shows the total power dissipation of a typical unloaded complementary flip-flop as a function of frequency or repetition rate.

A bistable IGFET circuit suitable for use as a storage cell in a random access memory is illustrated in Figure 12.19 [21]. Because of the low power drain, small area, and simple processing of these cells, large monolithic

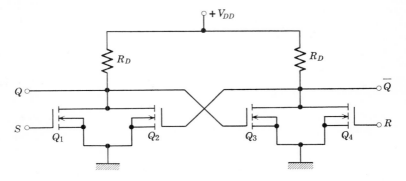

Figure 12.16 n-Channel IGFET set-reset flip-flop.

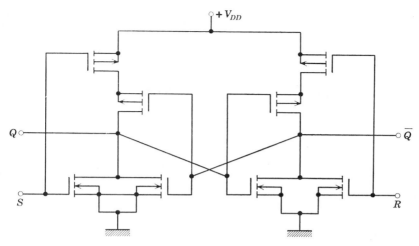

Figure 12.17 Complementary IGFET set-reset flip-flop.

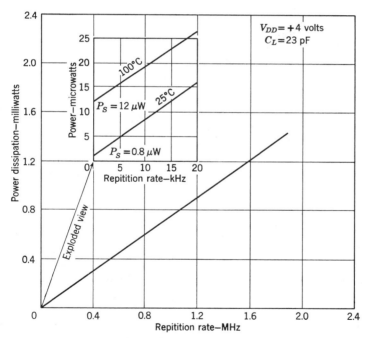

Figure 12.18 Power dissipation versus repetition rate for an unloaded complementary set-reset flip-flop (after Ahrons et al. [13]).

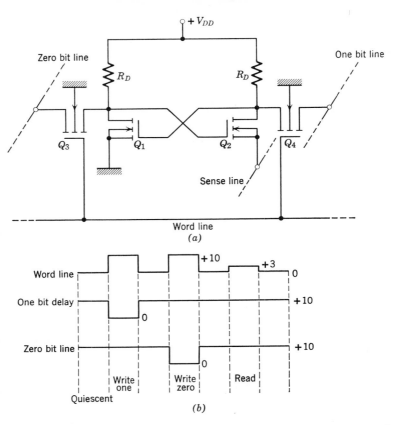

Figure 12.19 (a) n-Channel IGFET random access memory cell; (d) memory cell read and write pulse patterns.

memory arrays are possible [5,21–23]. Writing is accomplished in this cell by connecting, for example, the output node at the drain of Q_2 to 0 volts on the one bit line by turning on Q_4 with 10 volts on the word line. This turns off Q_1. If Q_1 is already OFF, no change of state occurs, but if Q_1 is ON, the regenerative action of the flip-flop causes it to change state. The nodes are charged or discharged through ON devices, resulting in short time constants and fast writing.

The process of reading occurs, for example, by partially turning on Q_3 and Q_4 with a 3 volt pulse on the word line, and detecting the presence of current flow through Q_2, which is ON. Turning on Q_3 and Q_4 connects the drain nodes to a 10 volt bit line voltage, which is greater than V_{DD}. If Q_2 is ON, current flows in the sense line indicating a stored one; if Q_2 is OFF, the relative absence of current flow in the sense line indicates a stored zero. Nondestructive reading is accomplished by ensuring that the maximum transient voltage on the ON node is below the threshold of the OFF IGFET.

It is apparent that the bidirectional characteristic of the IGFET that permits it to function as a high-quality transmission gate is quite useful in the operation of this storage cell. For other applications such as binary counting, transmission gate circuits are more efficient logic elements than standard Boolean gates [17,20]. For most monolithic designs, R_D in Figure 12.19 can be implemented with an enhancement IGFET. Complementary IGFET's can also be used in this application [15,22,23]. It is of particular importance that an IGFET memory cell be designed for micropower operation, in order to achieve the sizeable storage capacities, which are required in large digital systems without causing severe heat removal problems.

12.4 MULTIPHASE DYNAMIC CIRCUITS [7,25,26,24]

The previous sections of this chapter discuss the use of field effect transistors in circuit configurations that clearly emulate bipolar transistor digital circuits. The unique properties of an IGFET, including its high input impedance, inherent temporary memory of the gate-to-channel capacitance, and symmetrical output characteristics, provide an opportunity for digital circuit configurations that are not readily imitated with bipolar transistors. Typically, such IGFET circuit configurations operate in a dynamic mode that makes use of temporary memory, multiphase clocking, and switched load transistors to obtain extremely high element densities in monolithic integrated circuits as well as excellent micropower performance.

A layout comparison of a bipolar transistor and an IGFET or MOS integrated inverter for equal narrowest acceptable line width is illustrated in Figure 12.20 [24]. Using consistent layout criteria—a 200 ohms per square

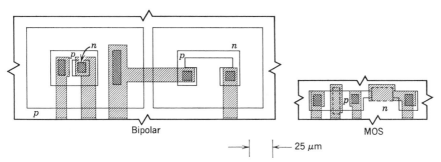

Figure 12.20 Layout comparison of integrated inverters using a bipolar transistor and a p-channel enhancement IGFET (after Warner [24]).

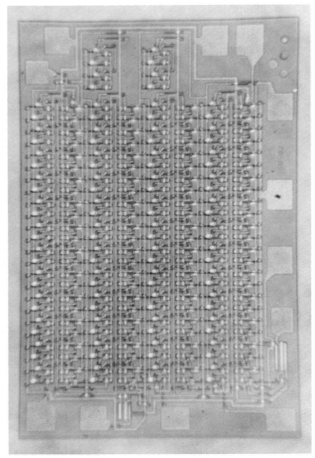

Figure 12.21 Microphotograph of a 90-stage MOS shift register containing over 500 transistors. Chip size is approximately 60 × 100 mils (courtesy of General Instruments, Inc.).

sheet resistance for the bipolar circuit and a 40,000 ohms per square sheet resistance for the IGFET load resistor—the area required by the bipolar circuit is 5.5 times larger than the IGFET circuit area. This important area advantage of the IGFET integrated circuit applies for more complex digital functions as well. Figure 12.21 illustrates the level of complexity that is feasible. The circuit is a 90-stage dynamic shift register that contains over 500 p-channel enhancement MOS transistors fabricated in a single silicon die.

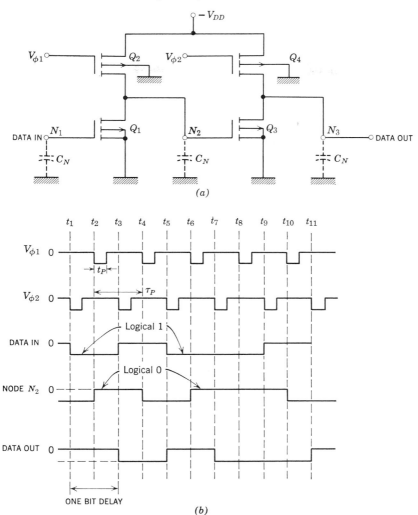

Figure 12.22 (a) Schematic diagram of an elementary two-phase dynamic logic stage; (b) waveform timing diagram for two-phase dynamic logic.

Two-Phase Circuits

The operation of a dynamic multiphase logic stage can be illustrated by Figure 12.22. At time t_1 a logical "1," corresponding to a voltage of approximately $-V_{DD}$, is stored on the capacitance C_N associated with the data input node N_1 or the gate of Q_1. At t_2 the phase-one clock voltage $V_{\phi 1}$ turns on Q_2 so that both Q_1 and Q_2 conduct and the potential of node N_2 rises to approximately 0 volts corresponding to a logical 0. At t_3 the phase-two clock voltage $V_{\phi 2}$ turns on Q_4, but since N_2 or the gate of Q_3 is near 0 volts, Q_3 does not conduct and C_N of node N_3 is charged through Q_4 to a potential near $-V_{DD}$ corresponding to the logical "1" state. Thus at t_3 the state of data OUT is the same as the state of data IN at t_1. The cycle time or bit delay time is equal to the interval from t_1 to t_3.

The dynamic logic circuit in Figure 12.22 consists of a delay stage plus control gating and has the logical properties of a Type D or delay flip-flop. It can be used as a recirculating shift-register stage and as the basis for clocked R-S and J-K flip-flops. When a load transistor is OFF, information is stored on the node capacitance C_N, which consists of the capacitance of a p-n junction, metal interconnections, and a metal gate. The upper limit on the time charge can be stored on the node capacitance as determined by the drain to substrate p-n junction of Q_1. Typically, minimum clock rates in the kilohertz range are required to ensure no loss of information. Maximum clock rates are determined by switching speed considerations, as already discussed.

The two-phase circuit in Figure 12.23 is an improved version of the circuit in Figure 12.22 in which the isolation between successive storage nodes is improved by the addition of transistors such as Q_2 and Q_5. The major disadvantage of this circuit is that considerable chip area is taken up by the two series inverter transistors Q_1 and Q_2 (or Q_4 and Q_5). To obtain essentially a zero output voltage at the storage node, as well as to reduce the time required to discharge a node capacitance, the combined resistance of Q_1 and Q_2 must be approximately the same as that of Q_1 in Figure 12.22. Therefore both Q_1

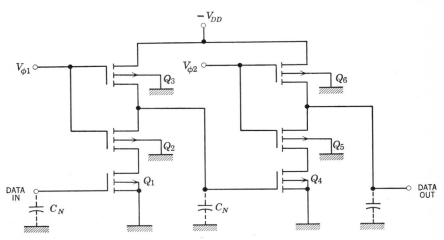

Figure 12.23 Two-phase dynamic logic stage with improved isolation between storage nodes.

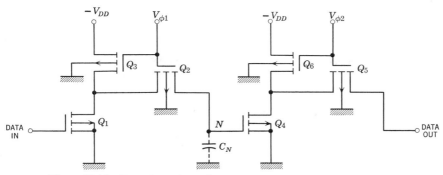

Figure 12.24 Two-phase dynamic logic stage using a transmission gate.

and Q_2 in Figure 12.23 must be larger in area than the single inverter transistor in Figure 12.22. In addition, both Q_1 and Q_2 must be about ten times larger than the load transistor Q_3 as discussed in connection with (12.21).

A further improvement in the two-phase circuit is illustrated in Figure 12.24. In this configuration Q_2 can be a minimum size device since the zero level does not depend upon its relative drain-to-source resistance in the ON state. Storage node isolation is enhanced by using Q_2 as a transmission gate. The direction of current flow through Q_2 reverses as the node capacitance C_N charges through Q_2 and Q_3 and discharges through Q_2 and Q_1. Ninety of these delay stages are fabricated in the silicon die pictured in Figure 12.21.

From the viewpoint of micropower operation, the salient feature of the two-phase circuits in Figures 12.22, 12.23, and 12.24 is that appreciable current is drawn from the power supply only during the clock pulse time t_p or when $V_{\phi 1}$ and $V_{\phi 2}$ are at their low level. Consequently the average power dissipation is a function of the clock duty cycle, t_p/τ_p, as described by Figure 12.22. The lower limit on t_p is determined by the switching speed of the circuit and $\tau_p > 2t_p$ is a requirement of the two-phase mode of operation. With t_p set at its minimum value, power dissipation is proportional to the clock frequency $1/\tau_p$. Consequently the average power drain can be reduced to extremely low values. In essence, the use of switched load resistors permits this reduction.

Four-Phase Circuits

Two prominent disadvantages that remain in the two-phase circuit in Figure 12.24 are (a) the requirement for dc "voltage divider" action between Q_1 and Q_3 (or Q_4 and Q_6) when both are conducting and thus the need for larger geometry for Q_1, and (b) the dc conducting path between $-V_{DD}$ and ground that exists when both Q_1 and Q_3 are conducting. The four-phase circuit in Figure 12.25 corrects these problems. The storage capacitance C_N at node N is precharged at phase-one ($\phi 1$) time when Q_3 is switched on. At $\phi 2$ time, Q_2 is switched on and C_N discharges through Q_2 and Q_1 if the input data is a logical one, or C_N remains charged if the input data is a logical zero. The same sequence of precharging and data sampling occurs for Q_4, Q_5, and Q_6 during $\phi 3$ and $\phi 4$ time so that the input data is shifted to the following stage. Current flows only when capacitances are charged or discharged as in the case of complementary IGFET logic. Thus the average power drain

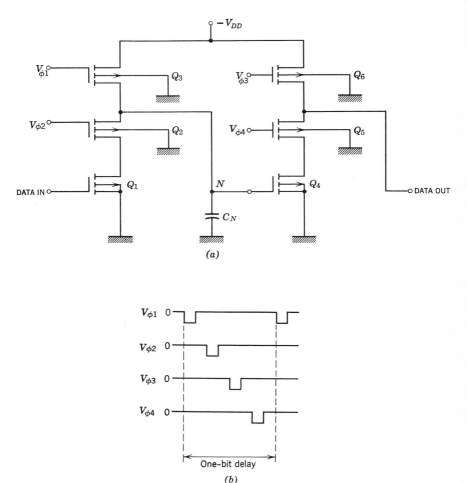

Figure 12.25 (a) Four-phase dynamic logic stage; (b) waveform timing diagram.

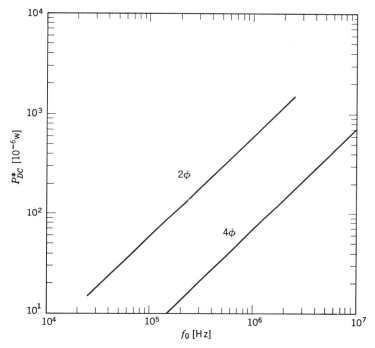

Figure 12.26 Typical average power dissipation versus frequency comparison for two-phase (Figure 12.24) and four-phase (Figure 12.25) dynamic logic states (after Farina [27]).

per node in this four-phase system is approximately

$$P_{DC}^{*} \simeq C_N V_{DD}^2 f_0, \qquad (12.32)$$

where $f_0 = 1/\tau_P$ is the clock frequency. Typical speed-power performance for multiphase circuits is illustrated in Figure 12.26 [27]. In order to extend the logical power of a four-phase (or a two-phase) system, inverter transistors such as Q_1 and Q_4 can be replaced by multiple input gates such as those discussed in Section 12.2. Further variations of the multiphase mode of operation of IGFET digital circuits have been reported [27].

REFERENCES

[1] A. K. Rapp, "Applications of Field Effect Transistors in Digital Circuits," in *Field-Effect Transistors: Physics, Technology and Applications*, Ed. Wallmark and Johnson, Prentice-Hall, Englewood Cliffs, N.J., 1966, pp. 312–358.

[2] A. W. Lo, *Introduction to Digital Electronics*, Addison-Wesley, Reading, Mass., 1967, Chapter 2.

[3] J. E. Price, "A Compatible MOS-Bipolar Device Technology for Low Power Integrated Circuits," IEEE Electron. Devices Meeting, Washington, D.C. (October 1966).
[4] R. D. Lohman, "Some Applications of Metal-Oxide Semiconductors to Switching Circuits," *Semiconduct. Prod. Solid-State Tech.*, **7**, 31–34 (May 1964).
[5] J. D. Schmidt, "Integrated MOS Transistor Random Access Memory," *Solid State Design*, **6**, 21–25 (January 1965).
[6] R. H. Norman, "Complex MOS Circuit Applications," *The NEREM Record*, Vol. VIII 172–173 November 1965.
[7] Engineering Staff of Philco-Ford, "MOS Monolithic Subsystems: A Revolution in Microelectronics," (1966).
[8] D. E. Farina and D. Trotter, "MOS Integrated Circuits Save Space and Money," *Electronics*, **38**, 84–95 (October 4, 1965).
[9] J. L. Seely, "Designing with MOS Semiconductors," General Instrument Co., Technical Bulletin.
[10] F. M. Wanlass and C. T. Sah, "Nanowatt Logic Using Field-Effect Metal-Oxide Semiconductor Triodes," *Inter. Solid-State Circuits Conf., Dig. Tech. Papers*, Vol. VI, 32–33 (February 1963).
[11] G. E. Moore et al., "Metal-Oxide-Semiconductor Field-Effect Devices for Micropower Logic Circuitry," in *Micropower Electronics*, E. Keonjian, Ed., Macmillan, 1964, pp. 41–55.
[12] J. R. Burns, "Switching Response of Complementary-Symmetry MOS Transistor Logic Circuits," *RCA Review*, **XXV**, 627–661 (December 1964).
[13] R. W. Ahrons et al., "MOS Micropower Complementary Transistor Logic," *Solid-State Circuits Conf., Dig. Tech. Papers*, Vol. VIII, 80–81 (February 1965).
[14] M. M. Mitchell and R. W. Ahrons, "MOS Micropower Complementary Transistor Logic," Session I, WESCON, San Francisco, California, August 1965.
[15] J. R. Burns et al., "Integrated Memory Using Complementary Field-Effect Transistors," *Digest of Technical Papers of International Solid-State Circuits Conference*, February 1966, pp. 116–119.
[16] K. K. Yagura et al., "Monolithic MOS Complementary Pairs," IEEE Electron Devices Meeting, Washington, D.C. (October 1966).
[17] A. K. Rapp et al., "Complementary-MOS Integrated Binary Counter," *Digest of Technical Papers of International Solid State Circuits Conference*, February 1967, pp. 52–53.
[18] M. H. White and J. R. Cricchi, "Complementary MOS Transistor," *Solid State Electronics*, **9**, 991–1008 (October 1966).
[19] J. Kane, "Switch Over to Field Effects," *Electronic Design 24*, October 25, 1966, pp. 54–60; November 8, 1966, pp. 72–79.
[20] R. W. Ahrons and P. D. Gardner, "Complementary MOS Integrated Circuits," *Proceedings of Integrated Circuits Seminar*, Stevens Institute of Technology, Hoboken, New Jersey (February 1, 1967).
[21] P. Pleshko and L. M. Terman, "An Investigation of the Potential of MOS Transistor Memories," *IEEE Trans. Electron. Computers*, **15** (No. 4), 423–427 (August 1966).
[22] R. Igarashi et al., "A 150 Nanosecond Associative Memory Using Integrated MOS Transistors," *Digest of Technical Papers of International Solid-State Circuits Conference*, February 1966, pp. 104–105.
[23] J. Wood and R. G. Ball, "The Use of Insulated-Gate Field-Effect Transistors in Digital Storage Systems," *Digest of Technical Papers of International Solid-State Circuits Conference*, February 1965, pp. 82–83.

[24] R. M. Warner, Jr., "Comparing MOS and Bipolar Integrated Circuits," *IEEE Spectrum*, **4** (No. 6), 50–58 (June 1967).
[25] J. Karp and E. de Atley, "Use Four-Phase MOS IC Logic," *Electronic Design 7*, April 1, 1967, pp. 62–66.
[26] J. L. Seeley, "Advances in the State-of-the-Art of MOS Device Technology," *Semicond. Prod. Solid-State Tech.*, **10**, 59–62 (March 1967).
[27] D. E. Farina, "Advanced MOS Circuit Techniques for LSI," University of Wisconsin Engineering Institute on Large Scale Integration (May 1967).

Chapter 13

Applications

Micropower circuits are of particular interest for an extremely wide variety of applications in which energy source capacity must be severely restricted. Battery-operated portable equipment, space electronics, and certain types of biomedical electronic devices represent perhaps the three most prominent categories of application of micropower circuits. The purpose of this chapter is to survey briefly the salient features of several applications of micropower circuits in these categories.

13.1 PORTABLE EQUIPMENT

The importance of portable electronic equipment for military and consumer applications has been expanding quite rapidly in recent years. Both of these trends can be expected to continue for some time.

One of the more representative micropower portable military equipments is the FM helmet radio receiver [1,2] pictured in Figure 13.1. Figure 13.2 illustrates a block diagram of this double conversion receiver and Table 13.1 lists a summary of its salient performance characteristics. In order to meet the reliability, size, weight, cost, and micropower requirements of this receiver, it incorporates seven silicon monolithic integrated circuits of five different types. Typical performance of each circuit is summarized in Table 13.2. The integrated circuits illustrated by Figures 3.17, 4.8, 5.8, 7.2, and 9.6

Portable Equipment 239

Figure 13.1 Helmet radio receiver (courtesy of U.S. Army Electronics Command).

were designed in the course of the development of this receiver. Selectivity at the signal frequency (50 MHz) and high intermediate frequency (10.7 MHz) is obtained by using discrete LC circuits while selectivity at the lower intermediate frequency (455 kHz) is provided by a ceramic filter. Besides improving operational effectiveness through prolonged battery life, the minimum

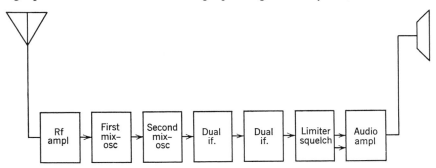

Figure 13.2 Simplified block diagram of Helmet Radio Receiver (after Quaid [2]).

Table 13.1 Receiver Characteristics

Frequency coverage	47–57 MHz
Sensitivity (open-circuit)	0.004 uuW
Power drain (squelched)	19.0 mW
Power drain ($P_0 = 5.0$ mW)	36.0 mW
Battery life (15:1 duty cycle)	100 hours
Temperature range	$-40°C$ to $+55°C$
Audio output power	5.0 mW
Audio distortion ($P_0 = 5$ mW)	4.0 %
Weight	8.1 oz

power design of this receiver tends to reduce its life cycle cost by curtailing battery replacement expenses.

In addition to the helmet radio receiver, many other types of portable military equipment include minimum power drain as a primary design constraint. Among these equipments are (a) multichannel radio transceivers including analog and digital frequency synthesizers, (b) secure communications devices such as vocoders, (c) message entry and readout devices, (d) surveillance devices such as hand-held radars, remote sensors, and infrared night vision binoculars, (e) electronic navigation devices such as manpack LORAN receivers, (f) IFF (information—friend or foe) devices, and (g) electronic countermeasures receivers and rescue beacons [3]. Battery-operated data processing equipment [4] for use in military applications such as missile control and checkout, command and control, target location, trajectory analysis, sound ranging, laser ranging, and communications message control and editing is an additional important area for utilization of micropower circuits. Fuse timers and programmers [5] for conventional artillery, mortars,

Table 13.2 Integrated Circuit Characteristics

Circuit	Chip Size (mils)	Power Drain (mW)	Power Gain (dB)
Rf amplifier	30 × 30	2.4	18
Mixer-oscillator	45 × 45	1.8	13 (1st mixer)
			20 (2nd mixer)
If amplifier	45 × 45	2.1	50
Limiter-squelch	55 × 55	3.9	15 (limiter)
			40 (sq ampls)
Audio amplifier	60 × 60	22.5 ($P_0 = 5$ mW)	40

mines, and booby traps represent a potentially massive application for micropower integrated circuits.

The portable transistor broadcast receiver is the most familiar consumer electronic device in which battery power drain is a significant design constraint. The electronic wristwatch is a consumer product in which micropower operation is extremely critical. Among the many other existing and potential portable electronic products for consumers, perhaps the most interesting from an economic standpoint is the portable television receiver. An experimental single channel micropower television receiver that can be carried in a shirt pocket has been demonstrated [6]. Two problems that must be solved in order to develop a practical version of this receiver are an increase in its battery life and a very low-power, compact planar display device.

13.2 SPACE ELECTRONICS

The successful design of virtually all electronic subsystems for satellites and space vehicles is critically dependent on minimizing average power consumption. Figure 13.3 illustrates the block diagram of one such subsystem—an analog-to-digital converter that was designed for micropower operation [7] in space applications.

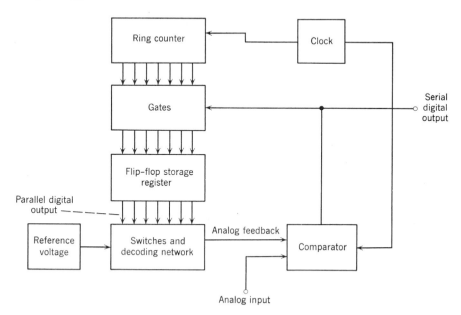

Figure 13.3 Block diagram of micropower analog-to-digital converter (after Sturman [7]).

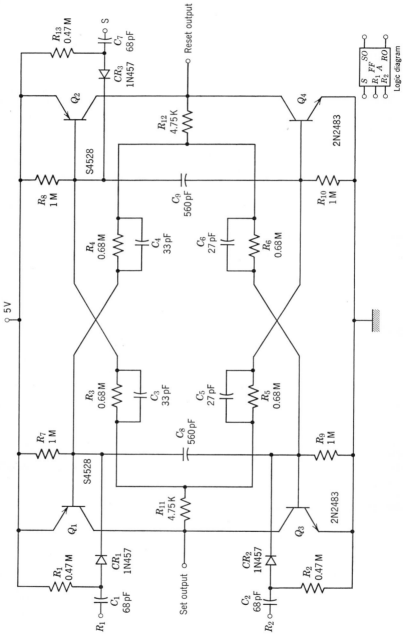

Figure 13.4 Bistable multivibrator used in flip-flop storage register of A- to D-converter (after Sturman [7]).

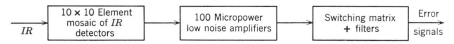

Figure 13.5 Simplified block diagram of IR search-track subsystem including micropower amplifiers.

This converter compares the analog input voltage with an analog feedback voltage generated by a predominantly digital system. The clock and ring counter provide a sequence of pulses to program the operation of the converter. These pulses are gated into a storage register that controls a digital-to-analog conversion network, which approximates the input as 1 of 128 discrete voltage levels. The analog feedback is generated by summing increments of $\frac{1}{2}, \frac{1}{4}, \frac{1}{8}$, and so on, of full scale in the decoding network; the feedback signal is compared to the input at each step. Whenever the total feedback signal exceeds the input, the comparator directs the storage register to remove the last increment added, and the system then proceeds to the next smaller increment, which is one-half as large. At the end of seven comparisons, the feedback signal matches the input to 1 part in 128, at which time the digital representation of the input is read out from the register.

Figure 13.4 illustrates the schematic diagram of the micropower flip-flop used in the storage register of the converter. The nominal power dissipation of the circuit is 80 μW. Using such circuits, the converter provides an absolute accuracy of ± 1 count for a total power drain of less than 10 milliwatts. With a full scale input of 20 millivolts converted into a seven-bit binary word 250 times per second, one count corresponds to 160 microvolts, which is quite suitable for a variety of space vehicle applications. The operating temperature range is $-20°C$ to $+80°C$. A generally similar converter has been fabricated utilizing monolithic integrated circuits [8].

An infrared search-track system that uses 100 low-noise micropower channel amplifiers excited by an infrared detector mosaic is illustrated in Figure 13.5 [9]. Details of the amplifier performance are given in Figure 6.7b.

13.3 BIOMEDICAL ELECTRONICS

Biomedical electronic devices that are implanted within the human body are among those whose power drain requirements are most stringent. The cardiac pacemaker is a prominent implantable device that supplies a periodic pulse of voltage to the heart to replace the defective natural stimulus and thereby restores normal cardiac rate. Figure 13.6 illustrates the schematic diagram of an implantable pacemaker circuit. The heart impedance is approximately equivalent to a series network of a 20-μf capacitance and

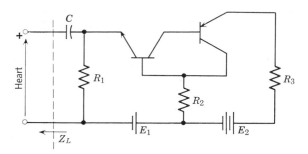

Figure 13.6 Implantable pacemaker circuit.

300 Ω resistance. When the capacitor C is completely discharged, the transistor configuration becomes regenerative and acts as a closed switch. The capacitor C is charged from batteries E_1 and E_2 with a time-constant given by the product of C and the sum of the load resistance R_L, transistor-saturation resistance, and R_3. The corresponding current surge constitutes the impulse imparted to the heart. When the capacitor voltage reaches the battery voltage (E_1 and E_2), the switch turns off and the transistors cease to conduct current. The capacitor C now discharges through R_1 and the load until the charge in C is almost nil, whereupon a new current surge generates another impulse. Some pertinent performance data for two versions of the circuit using 1.4 V mercury cells are listed [11]:

	4 cells	5 cells
current drain (dc)	14 μa	25 μa
power drain (dc)	82 μW	184 μW
pulse energy	23 μjoules	64 μjoules
power efficiency	30 %	38 %
expected battery life:		
rated capacity—1000 ma-hr	71,000 hrs	40,000 hrs
packaged volume	2.4 in³	4 in³

Biomedical telemetry is a growing area in which micropower design techniques are quite critical. Biomedical telemetry devices have enabled researchers to gather valuable information about the behavior of men, as well as many types of animals, birds, and fish. Figure 13.7 illustrates a six-channel FM/FM biotelemetry transmitter [12]. Designed for external surface mounting, the system has been used on a human subject to monitor two surface temperatures, one internal temperature, respiration rate, patient orientation, and muscle spasm. The system uses resistive transducers to measure the physiological phenomena. These transducers are powered in series by a temperature-compensated current source. Six subcarriers are linearly mixed and fed to the

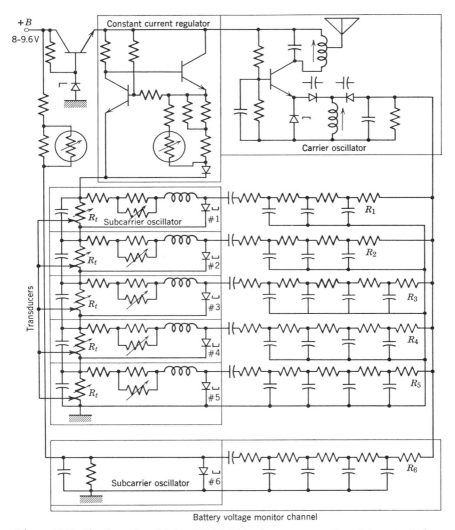

Figure 13.7 Six-channel multiplex FM/FM biotelemetry transmitter (after Ko [12]).

tunnel diode *RF* carrier oscillator. Resistances, R_1 through R_6, are carefully selected to prevent interchannel interference; *RC* filters are used to eliminate harmonics in the tunnel diode subcarrier oscillators. The transmission system occupies a 0.6 × 2.5 × 2.5 inch volume, weights 15 grams without the battery pack, draws 7 ma at the nominal voltage of 8.8 volts, and transmits at 110 to 130 MHz.

REFERENCES

[1] R. A. Gilson and T. B. Quaid, "An Integrated Circuit Helmet Radio Receiver," IEEE National Convention, New York (March 1967).
[2] T. B. Quaid, "New Dimensions in Military Radios," *Eng. Bull.*, Motorola Government Electronics Division, **15,** 16–25 (1967).
[3] L. F. Wagner and J. D. Meindl, "Static and Dynamic Performance of Micropower Transistor Logic," WESCON, San Francisco, Calif. (August 1965).
[4] E. Lieblein et al., "Minipac—A Battery Powered, Low Cost Field Computer," National Convention on Military Electronics, Washington, D.C. (September 1963).
[5] D. N. Shaw, "A Description of the Use of Semiconductor Integrated Circuits for Artillery Timers and Other Ordnance Material," IEEE Winter Convention on Military Electronics, Los Angeles, Calif. (February 1964).
[6] D. J. Tanner, "An Experimental Shirt-Pocket TV Set," *IEEE Trans. Broadcast and Television Receivers*, **12,** 141–146 (July 1966).
[7] J. C. Sturman and A. V. Bertolino, "Low-Power, Low-Level Analog-to-Digital Converter for Space Vehicle Applications," NASA TN D-2916 (July 1965).
[8] W. W. Gaertner et al., "Minimum-Power Microelectronic Space Systems," *IEEE Trans. Aerospace*, **2,** 241–251 (April 1964).
[9] C. P. Hoffman and M. N. Giuliano, "Micropower Molecular Amplifiers for Infrared Search-Track System Application," *Solid State Design*, **6,** 24–27 (March 1965).
[10] H. Raillard, "Development of an Implantable Cardiac Pacemaker," *1962 Inter. Solid-State Circuits Conf., Dig. Tech. Papers*, Philadelphia, Penn. (February 1962).
[11] Wilson Greatbatch et al., "Implantable Cardiac Pacemakers," IEEE International Convention, New York (March 1965).
[12] W. H. Ko and L. E. Slater, "Bioengineering: A New Discipline," *Electronics*, Vol. 38, pp. 111–118 (June 14, 1965).

List of Symbols

		Chapter
a	basic amplifier gain of elementary feedback amplifier	(8)
a	transformer turns ratio	(3), (5), (9)
a_c	half-thickness of channel of JGFET	(1)
A_{dc}	ratio of incremental differential-mode output voltage to common-mode input voltage	(10)
A_{dd}	ratio of incremental differential-mode output voltage to differential-mode input voltage	(10)
A_{ic}	conversion current gain of AM detector	(7)
A_{i0}	midband current amplification per stage	(3)
A_{i0}^*	midband current amplification of overall amplifier	(3)
A_j	area of p-n junction	(1)
a_0	basic amplifier midband gain of elementary feedback amplifier	(8)
a_t	transition region impurity concentration gradient of a linearly graded p-n junction	(1)
A_v	complex voltage amplification of an amplifier stage	(4)
A_v^*	overall complex voltage amplification of a wideband amplifier	(4)
A_v'	voltage gain	(8)
A_{vc}	conversion voltage gain of mixer	(7)
A_{v0}	midband voltage amplification of a wideband amplifier stage	(3), (4)
A_{v0}^*	midband voltage amplification of overall amplifier	(3)

247

Applications

A_{vp}	voltage gain of alternate pair	(8)
B_0	output susceptance of harmonic oscillator	(9)
B_T	susceptance of tuned circuit in parallel with harmonic oscillator load conductance	(9)
C	capacitance	(1)
C_B	bypass capacitance	(3)
C_{BE}	average base-to-emitter capacitance for a specified range of V_{BE}	(11)
C_{be}	transistor emitter transition region plus diffusion capacitance	(1)
C_c	transistor collector transition region plus diffusion capacitance	(1)
C_{c1}	transistor collector-to-base capacitance directly under the emitter	(1)
C_{c1}	input coupling capacitance	(3)
C_{c2}	transistor collector-to-base capacitance not directly under the emitter	(1)
C_{CB}	average collector-to-base capacitance for a specified range of C_{CB}	(11)
C_d	minority carrier diffusion capacitance outside the transition region of p-n junction	(1)
C_{ds}	drain-to-source capacitance of FET	(1), (4)
C_E	emitter resistor bypass capacitance	(3)
C_{gc}	gate-to-channel capacitance of FET	(1)
C_{GD}	average gate-to-drain capacitance for a specified range of V_{GD} and V_{GS}	(12)
C_{gd}	gate-to-drain capacitance of FET	(1), (4)
C_{GS}	average gate-to-source capacitance for a specified range of G_{GS} and V_{GD}	(12)
C_{gs}	gate-to-source capacitance of FET	(1), (4)
C_j	total capacitance of a p-n junction	(1)
C_{jt}	tunnel-diode junction capacitance	(1)
C_L	equivalent load capacitance	(4), (11), (12)
C_L	load coupling capacitance	(3)
$CMRR$	common-mode rejection ratio	(10)
C_N	storage node capacitance of FET circuit	(12)
C_s	stray capacitance	(1)
C_T	tuning capacitance	(5), (9)
C_{tc}	transition region carrier capacitance	(1)
C_{ti}	transition region space charge capacitance	(1)
D_n, D_p	diffusion constant for electrons and holes, respectively	(1)

List of Symbols

E	total energy dissipated in charging and discharging a capacitor	(12)
$\overline{e_b^2}$	mean square base resistance noise voltage of transistor	(6)
$\overline{e_{eq}^2}$	mean square equivalent noise voltage of a transistor	(6)
$\overline{e_s^2}$	mean square source noise voltage	(6)
E_s	rms source voltage	(3)
E_s	amplitude of base-to-emitter signal voltage of AM detector with no modulation present	(7)
$e_s(t)$	instantaneous base-to-emitter signal voltage on AM detector	(7)
F	noise figure	(7)
F	reverse transfer function of feedback network of amplifier with multistage feedback loop	(8)
f	reverse transfer function of feedback network of elementary feedback amplifier	(8)
f_β	frequency where $|h_{fe}| = h_{fe0}/\sqrt{2}$	(4)
f_c	transconductance cutoff frequency of FET	(1)
f_{cl}	lower noise corner frequency of transistor	(6)
f_{ch}	upper noise corner frequency of amplifier	(6)
f_{ch}	upper noise corner frequency of transistor	(6)
f_{cr}	resistor cutoff frequency	(1)
f_{ct}	junction cutoff frequency of tunnel diode	(1)
F_f	noise figure in flicker noise region	(6)
$f_{\max}$	maximum frequency of oscillation of tunnel diode	(1)
F_0	noise figure in shot noise region	(6)
f_0	frequency of operation or repetition rate of an FET digital circuit	(12)
f_0 (max)	maximum frequency of oscillation of a harmonic oscillator	(9)
f_s	signal frequency	(6)
f_T	frequency where $|h_{fe}| = 1$	(4)
f_T	transistor current gain-bandwidth product	(1)
g	small-signal conductance	(1)
$G_1, G_2, \ldots, G_N$	power gain of the first, second, ..., nth stage of an amplifier	(3)
G_B	base bias conductance ($1/R_B$)	(3)
G_C	collector load conductance ($1/R_C$)	(3)

250 Applications

g_d	drain transconductance of FET in triode region	(1)
G_L	harmonic oscillator load conductance	(9)
G_L	ac load conductance ($1/R_L$)	(3)
g_m	forward transconductance of bipolar and field effect transistors	(1)
g_{m1}	conversion transconductance of mixer	(7)
g_{m0}	transconductance of JGFET for $V_G = 0$ in pinch-off region	(1), (4)
G_0	output conductance of harmonic oscillator	(9)
G_P	actual power gain	(5)
G_{P0}	actual power gain at center frequency	(5)
g_s	source transconductance of JGFET	(1)
G_T	transducer power gain	(3)
h_{FE}	$I_C/(I_B + I_{CB0})$, the static base-to-collector current gain	(2)
h_{fe0}	transistor small-signal low-frequency current gain	(1)
I	direct current	(2)
$\overline{i_a^2}$	mean square collector junction reverse current noise of transistor	(6)
$\overline{i_b^2}$	mean square base current noise of transistor	(6)
$\overline{i_c^2}$	mean square collector current noise of transistor	(6)
$\overline{i_f^2}$	mean square flicker noise current of transistor	(6)
I_B	static transistor base current	(1)
$i_{be}(t)$	instantaneous current through the transistor base diffusion resistance r_{be}	(7)
I_{B0}	static base current of a cutoff transistor	(11)
I_{BS}	static base current of a saturated transistor	(11)
I_C	static transistor collector current	(1)
$i_C(t)$	instantaneous collector current	(7)
$I_c(\omega_m)$	rms value of collector current of AM detector at modulation frequency	(7)
I_{CB0}	ideal saturation current of collector junction with $I_E = 0$	(1)
I_{C0}	static collector current of a cutoff transistor	(11)
I_{CQ}	collector current of mixer with zero local oscillator injection voltage	(7)

List of Symbols

Symbol	Description	Ref
I_{CQ}	quiescent collector current of AM detector	(7)
I_{CQ}	quiescent collector current of harmonic oscillator	(9)
I_{CBS}	reverse saturation current of collector junction with $V_{BE} = 0$	(1), (3)
I_{CS}	static collector current of a saturated transistor	(11)
I_D	drain current of FET	(1), (4)
i_D	instantaneous drain current of FET	(12)
I_{DSS}	static FET drain current with zero gate-to-source voltage	(1), (4), (12)
I_E	static transistor emitter current	(1)
I_{EB0}	ideal saturation current of emitter junction with $I_C = 0$	(1)
I_{EBS}	ideal saturation current of emitter junction with $V_{CB} = 0$	(1)
I_{EX}, I_{CX}	coefficient of the nonideal generation-recombination currents for the emitter and collection junctions, respectively	(1)
I_{in}	input current of wideband amplifier	(4)
I_J	coefficient of the total current of a non-ideal p-n junction	(1)
I_L	inverter static load current	(11)
I_L	peak ac load current	(3)
i_L	inverter instantaneous load current	(11)
I_m	Miller effect current of an iterative wideband amplifier stage	(4)
I_0	minimum permissible dynamic collector current	(3)
I_p	peak current of tunnel-diode	(1)
I_S	reverse saturation current of ideal p-n junction	(1)
$I_s(\omega_s)$	rms value of signal source current of AM detector	(7)
$I_{S1}, I_{S2}, I_{S3}, I_{SC}$	reverse isolation junction currents of diffused silicon resistors	(2)
I_{ST}	reverse isolation junction current of a monolithic transistor	(2)
I_V	valley-current of tunnel-diode	(1)
I_X	coefficient of the nonideal generation-recombination currents in a p-n junction	(1)
K	stability factor	(5)
k	Boltzmann constant	(1)

K_f	flicker noise constant of transistor	(6)
L_b	diffusion length of minority carrier in base region	(10)
L_c	channel length of FET	(1)
L_n, L_p	diffusion length for electrons and holes, respectively	(1)
L_r	resistor length	(1)
L_s	tunnel diode series inductance	(1)
L_T	tuning inductance	(5)
L_T	inductance in parallel with load conductance of harmonic oscillator	(9)
M	fan-in	(11)
m	index of modulation for AM detector input	(7)
N	fan-out	(11)
n	junction ideality factor ($n = 1$ for ideal junction)	(1)
n	number of stages	(3), (4)
n_i	carrier concentration in intrinsic semiconductor material	(1)
n_{p0}	electron density at p-n junction boundary in p region	(1)
$P_1, P_2, \ldots, P_N$	ac power into first, second, ..., nth stage of an amplifier	(3)
P_{AVS}	ac power available from source	(2)
P_{DC}	quiescent power of a single stage	(3), (4)
P_{DC}^*	total quiescent power drain of circuit	(3), (4)
P_{DC}'	normalized quiescent power	(4)
P_L	ac load power	(3)
p_{n0}	hole density at p-n junction boundary in n region	(1)
P_S	standby power drain of FET circuit	(12)
Q	quality factor of a tuned circuit	(5)
Q	quiescent point	(2)
q	electronic charge	(1)
R	resistance	(1), (2)
R_1	external base padding resistance of an RTL gate	(11)
R_1, R_2, R_3	dc base bias circuit resistances	(2)
$R_1, R_2, \ldots, R_N$	input resistance of the first, second, ..., nth stage of an amplifier	(3)
R_B	external base resistance	(11)
R_B	equivalent ac resistance of bias circuit resistors	(2), (3)

List of Symbols

R_{B1}, R_{B2}, R_{B3}	base bias network resistances	(3)
R_b	equivalent ac resistance of base bias network	(3)
r_b	transistor base resistance	(1)
R_{BB}	base bias resistance	(3)
r_{be}	transistor small-signal common-emitter diffusion resistance	(1)
R_C	dc collector load resistance	(2), (11)
r_c	effective channel resistance of FET	(1)
r_c	transistor collector-to-base resistance caused by base-width modulation	(1)
R_{CB}	collector resistance	(3)
r_{ce}	transistor collector-to-emitter resistance resistance caused by base-width modulation	(1)
R_D	dc load resistor for an FET	(3), (4), (12)
r_{dd}	series drain resistance of FET	(1)
r_{ds}	drain-to-source resistance of FET	(1)
R_E	dc emitter degeneration resistance	(2)
R_e	ac emitter degeneration resistance	(3), (8)
R_{eq}	equivalent noise resistance of transistor	(6)
R_F	dc feedback resistance in feedback pair circuit	(2)
R_f	collector feedback resistance	(8)
r_{gd}	gate-to-drain resistance of FET	(1)
r_{gs}	gate-to-source resistance of FET	(1)
R_{ie}	input resistance of wideband amplifier	(4)
r_{jt}	tunnel-diode junction resistance	(1)
R_L	equivalent ac load resistance	(4)
R_L	equivalent load resistance of inverter	(11)
R_N	input node resistance of DTL and TTL circuits	(11)
R_0	effective load resistance of mixer at the output frequency	(7)
R_p	resistance in parallel with transistor at input of wideband amplifier	(4)
R_s	source resistance	(3), (6)
r_s	tunnel-diode series bulk resistance	(1)
r_{ss}	series source resistance of FET	(1)
R_T	base turn-off resistance	(11)
R_T	total resistance per stage	(3)
R_t	tuned circuit equivalent shunt resistance	(7)

Applications

R_T^*	total resistance of amplifier	(3)
S	sensitivity of overall amplifier gain to changes in transistor current gain	(8)
s	complex frequency, $\sigma + j\omega$	(8)
T	absolute temperature	(1), (3)
T	loop gain of elementary feedband amplifier	(8)
t	time	(3)
t_d	turn-on delay time of a transistor inverter	(11)
t_f	fall time of a transistor inverter	(11)
t_{f1}	fall time interval of inverter during which transistor is active	(11)
t_{f2}	fall time interval of inverter during which transistor is cutoff	(11)
T_n	nominal operating temperature of circuit	(3)
T_0	290°K	(6)
T_{0x}	thickness of gate insulator of IGFET	(1)
t_{pd}	average propagation delay time	(11)
t_r	rise time of a transistor inverter	(11)
t_s	storage time of a transistor inverter	(11)
t_{sw}	total switching time of an inverter	(11)
T_x	upper operating temperature limit	(2), (3)
T_y	lower operating temperature limit	(2), (3)
V	dc voltage	(2)
$V_{\phi 1}, (V_{\phi 2}, V_{\phi 3}, V_{\phi 4})$	phase-one (-two, -three, -four) clock voltage of multiphase FET circuit	(12)
V_B	inverter static input voltage level	(11)
V_B	the equivalent bias circuit supply voltage	(2)
v_B	inverter instantaneous input voltage level	(11)
V_{BB}	turn-off or negative supply voltage for a logic gate	(11)
V_{BE}	static transistor base-to-emitter voltage	(1)
$v_{be}(t)$	instantaneous base-to-emitter voltage of transistor	(7)
V_{BE0}	static base-to-emitter voltage of a cutoff transistor	(11)
V_{BES}	static base-to-emitter voltage of a saturated transistor	(11)
V_{BET}	threshold base-to-emitter voltage at which a cutoff transistor begins to turn on	(11)
V_{CB}	static transistor collector-to-base voltage	(1)
V_{cb}	amplitude of common base harmonic oscillator output voltage	(9)
V_{CC}	collector supply voltage	(2), (4)

List of Symbols

V_{CE}	static transistor collector-to-emitter voltage	(1)
V_{ce}	amplitude of common-emitter harmonic oscillator output voltage	(9)
V_{CE0}	static collector-to-emitter voltage of a cut-off transistor	(11)
V_{CES}	static collector-to-emitter voltage of a saturated transistor	(11)
V_{CEX}	collector-to-emitter voltage level at which inverter transistor passes from active to cutoff region during fall time interval	(11)
V_D	dc diode voltage drop	(2)
V_{DD}	dc drain supply voltage of FET	(12)
V_{DS}	static drain-to-source voltage of FET	(1), (4), (12)
v_{DS}	instantaneous drain-to-source voltage of FET	(12)
V_{DST}	static drain-to-source voltage of FET operating in triode region below pinch-off	(12)
V_E	dc emitter feedback voltage	(3)
V_e	ac emitter feedback voltage	(3)
V_{gc}	small-signal gate-to-channel voltage of FET	(1)
V_{GG}	dc gate supply voltage of FET	(12)
V_{GS}	static gate-to-source voltage of FET	(1), (4), (12)
V_{GST}	static gate-to-source voltage operating in triode region below pinch-off	(12)
V_i	ac input voltage	(5)
V_{in}	input voltage of wideband amplifier	(4)
V_l	amplitude of local oscillator injection voltage	(7)
V_L	peak ac load voltage	(3), (4)
$v_l(t)$	instantaneous local oscillator injection voltage of mixer	(7)
V_0	ac output voltage	(5)
V_0	minimum permissable dynamic collector voltage	(3)
$V_0(\omega_0)$	rms value of mixer output voltage at the output frequency	(7)
V_P	pinch-off or threshold voltage of FET	(1), (4), (12)
V_P	voltage of peak current of tunnel-diode	(1)

Applications

Symbol	Description	Ref
V_S	source voltage of FET	(1)
V_s	amplitude of signal voltage input of mixer	(7)
$v_s(t)$	instantaneous signal voltage input of mixer	(7)
V_V	voltage of valley current of tunnel-diode	(1)
W	electrical base width	(10)
W_c	channel width of FET	(1)
W_r	resistor width	(1)
W_t	width of the transition region of a *p-n* junction	(1)
Y_i	input admittance	(5)
y_{12s}	stray reverse transadmittance of cascode circuit	(5)
Y_0	output admittance of harmonic oscillator	(9)
Y_S	source admittance	(5)
Z_F	feedback network impedance of harmonic oscillator	(9)
Z_{ie}	input impedance of wideband amplifier	(4)
Z_L	equivalent load impedance	(4)
Z_m	Miller effect impedance for an iterative wideband amplifier stage	(4)
Z_0	output impedance	(4)
α_N	ideal transistor static emitter-to-collector current transfer ratio	(1)
α_I	ideal transistor static collector-to-emitter current transfer ratio	(1)
γ	kT/q	(3), (4), (7)
γ_R	R_x/R_y ratio of the value of a resistor at its upper operating temperature limit to the value at its lower operating temperature limit	(2)
δ_R	% tolerance of a transistor expressed as a decimal	(2)
ΔA_V	difference in voltage amplification	(8)
Δf	narrow frequency interval or bandwidth	(6)
ΔV_0	dc stability margin or noise immunity of a cutoff transistor gate	(11)
ΔV_0	static noise margin of a cutoff FET gate	(12)
ΔV_S	dc stability or noise immunity of a saturated transistor gate	(11)
ΔV_S	static noise margin of a conducting FET gate	(12)

List of Symbols

Symbol	Description	Ref
$\Delta\omega$	bandwidth of a tuned amplifier stage	(5)
$(\Delta\omega)^*$	overall bandwidth of a multistage tuned amplifier	(5)
ϵ	base of natural logarithms = 2.7132	(3)
ϵ	dielectric constant	(1)
ϵ_{0x}	dielectric constant of gate insulator of IGFET	(1)
η	dc to ac efficiency	(3), (4), (9)
θ	overdrive factor for static base current of a saturated transistor	(11)
θ	R_B/R_C, ratio of ac base to ac collector resistance in low-frequency amplifier	(3)
μ_c	carrier mobility in channel of FET	(1)
ρ_s	sheet resistivity in ohms per square	(1)
σ	conductivity	(1)
τ_0	effective carrier lifetime within a reversed biased depletion region	(1)
φ_0	energy band gap of semiconductor material at 0°K	(10)
ω	angular frequency	(3)
ω_c	cutoff frequency of a tuned amplifier stage	(5)
ω_c	upper 3 dB cutoff frequency or bandwidth of a wideband amplifier stage	(4)
ω_c^*	overall upper 3 dB cutoff frequency or bandwidth of a wideband amplifier	(4)
ω_{ie}	bandwidth of input impedance of wideband amplifier	(4)
ω_l	angular frequency of oscillator	(7)
ω_m	angular frequency of modulation or output frequency of AM detector	(7)
ω_0	angular intermediate or output frequency of mixer	(7)
ω_0	center frequency or resonant frequency of a tuned amplifier stage	(5)
ω_0	frequency of signal voltage input of mixer	(9)
ω_s	angular frequency of signal voltage output of mixer	(7)
ω_s	angular signal frequency input of AM detector	(7)

Index

Amplifiers, direct current, 145–156
  common mode rejection, 147–149
  discrimination factor, 146
  drift, 151–152
  gain, 145–146
  input offset voltage, 150–151
  monolithic, 154–156
  noise, 152–153
 feedback amplifiers, 121–133
  bandwidth, 131
  distortion, 130
  local collector feedback, 126–129
  local emitter feedback, 123–128
  loop gain, 122
  multistage feedback, 129
  quiescent power, 133
  sensitivity, 121–123
  stability, 132
 low frequency, 41–66
  class A, 41–55
  class B, 55–57
  class D, 60–61
  complementary symmetry, 59
  driver states, 50–55
  input stages, 41–44
  intermediate stages, 45–48
  output stages, 56–60
 low noise, 98–109
  flicker noise range, 101
  low frequency noise range, 101–103
  monolithic, 106
  noise figure, 99–100
  shot noise range, 101–103
  transistor noise model, 98–99
 tuned, 84–97
  bandwidth, 87
  cascode, 92–94
  center frequency, 86
  gain, 85–86
  iterative stage, 85–89
  monolithic, 96
  multistage, 89–92
  stability, 88–89
 wideband, 67–83
  bandwidth, 71–72
  field effect transistor, 77–79
  gain, 71
  input stage, 67–68
  number of stages, 71–73
  output stages, 82

Biomedical electronics, 243–245

Capacitors, 22–24
Consumer electronics, 241

DC operating point, 29–40
 DC collector degeneration, 32–34
 DC emitter degeneration, 29–31
 diode compensation, 34

Index

transistor compensation, 35
Detectors, 116–120
  amplitude modulation, 116–120
  conversion gain, 119
  equivalent circuit, 116–117
  power drain, 119
Device models, 1–28
Digital circuits, 157–237
  bipolar transistor, 157–209
  field effect transistor, 210–237

Inverter, bipolar transistor, 157–169
  dynamic behavior, 161–169
  fall time, 164–165
  iterative chain, 169–172
  propagation delay time, 171–172
  rise time, 162–163
  static behavior, 157–161
  storage time, 163–164
  turn-on delay time, 161–162
  field effect transistor, 210–223

Junction, diodes, 1–3
  capacitance, 2–3
  diffusion current, 1–2
  excess current, 1–2

Logic gates, 173–207
  complementary, 196–201
  diode transistor (DTL), 187–189
  emitter coupled (ECL), 190–193
  field effect transistor, 223–226
  resistor transistor (RTL), 173–183
  transistor resistor (TRL), 186–187
  transistor transistor (TTL), 189–190
  tunnel device (TDL), 193–195
  worst case design, 173–183

Military electronics, 238–241
Mixers, 110–116
  conversion gain, 116
  equivalent circuit, 110–111
  monolithic, 116
  noise figure, 113–115
  quiescent power, 115
Multiphase dynamic circuits, 229–235

Operational amplifier, 154–156
Oscillators, harmonic, 134–144
  generalized design, 140–143
  monolithic, 144
  power drain, 136–139
  simplified design, 134–139

Parasitics, 26
Portable equipment, 238–241
Pulse-powered digital circuits, 201–207

Quiescent point, 29–40; *see also* DC operating point

Resistors, 22–25

Space electronics, 241–243
Storage circuits, 183–186, 226–229
  bipolar transistor, 183–186
  field effect transistor, 226–229

Transistors, 3–20
  bipolar, 3–13
    current gain, 6
    gain bandwidth product, 11
    large signal model, 3–9
    lumped model, 7–9
    small signal model, 10–13
  field effect, 14–20
    insulated gate, 18–20
    junction gate, 14–18
Tunnel diodes, 21–22

LIBRARY
BURROUGHS CORP.